Spectroscopy for Amateur Astronomers
Recording, Processing, Analysis and Interpretation

This accessible guide presents the astrophysical concepts behind astronomical spectroscopy, covering both the theory and the practical elements of recording, processing, analyzing and interpreting your spectra. It covers astronomical objects, such as stars, planets, nebulae, novae, supernovae, and events such as eclipses and comet passages. Suitable for anyone with only a little background knowledge and access to amateur-level equipment, the guide's many illustrations, sketches and figures will help you understand and practice this scientifically important and growing field of amateur astronomy, up to the level of pro–am collaborations. Accessible to non-academics, it benefits many groups, from novices and learners in astronomy clubs, to advanced students and teachers of astrophysics. This volume is the perfect companion to the *Spectral Atlas for Amateur Astronomers*, which provides detailed commented spectral profiles of more than 100 astronomical objects.

Marc F.M. Trypsteen is a Belgian pharmacist and astronomer, with a background in analytical chemistry and spectroscopy. He has lectured on astronomical spectroscopy at university, high schools, observatories and astronomy clubs. He reviewed the *Spectral Atlas* in detail and co-authored *Spectroscopy for Amateur Astronomers*. He is co-founder of the Astro Event Group, Belgium and also of the Astropolis Space Science Center in Ostend, where, in addition to outreach activities, he is responsible for the section on spectroscopy education and research.

Richard Walker has been an avid astronomer since he was aged 12. He spent his career in civil engineering, planning large projects such as power plants, dams and tunnels. Now retired, in the last 10 years he has focused increasingly on stellar astronomy and on the indispensable key to this topic – spectroscopy. He undertook a large observing project to record and document the spectra of the most important astronomical objects, and chose to share this gathered information for the benefit of other amateurs worldwide. The *Spectral Atlas* and *Spectroscopy for Amateur Astronomers* are the fruits of his labor. He lives near Zürich, in Switzerland.

Spectroscopy for Amateur Astronomers

Recording, Processing, Analysis and Interpretation

Marc F. M. Trypsteen
Amateur astronomer, Belgium

Richard Walker
Amateur astronomer, Switzerland

CAMBRIDGE
UNIVERSITY PRESS

CAMBRIDGE
UNIVERSITY PRESS

University Printing House, Cambridge CB2 8BS, United Kingdom

One Liberty Plaza, 20th Floor, New York, NY 10006, USA

477 Williamstown Road, Port Melbourne, VIC 3207, Australia

314-321, 3rd Floor, Plot 3, Splendor Forum, Jasola District Centre, New Delhi - 110025, India

79 Anson Road, #06-04/06, Singapore 079906

Cambridge University Press is part of the University of Cambridge.

It furthers the University's mission by disseminating knowledge in the pursuit of education, learning and research at the highest international levels of excellence.

www.cambridge.org
Information on this title: www.cambridge.org/9781107166189

First published 2017

A catalogue record for this publication is available from the British Library

Library of Congress Cataloging in Publication data
Names: Trypsteen, Marc F. M. | Walker, Richard, 1951-
Title: Spectroscopy for amateur astronomers : recording, processing, analysis, and interpretation / Marc F.M. Trypsteen, amateur astronomer, Belgium, Richard Walker, amateur astronomer, Switzerland.
Description: Cambridge : Cambridge University Press, 2017. | Includes bibliographical references and index.
Identifiers: LCCN 2016036436 | ISBN 9781107166189 (Hardback : alk. paper)
Subjects: LCSH: Stars–Spectra. | Spectrum analysis. | Amateur astronomy.
Classification: LCC QB871 .T79 2017 | DDC 522/.67–dc23
LC record available at https://lccn.loc.gov/2016036436

ISBN 978-1-107-16618-9 Hardback
9781316642566 – 2 volume Hardback set
9781107165908 – Volume 1
9781107166189 – Volume 2

CONTENTS

8

Calibration of the Spectra   67

9

Analysis of the Spectra   76

10

Temperature and Luminosity   85

15

Amateurs and Astronomical Science

PREFACE

Spectroscopy is the main key to astrophysics. Without it, not only our current picture of the Universe but also the actual understanding of the atomic structure and quantum mechanics would be unthinkable. Therefore, most of the larger professional optical observatories use a significant part of the telescope time for the recording of spectra. Over recent years technological advances like CCD cameras, and also affordable spectrographs, have come on the market, causing a significant, but surprisingly late, upturn of this crucial and fascinating discipline within the community of amateur astronomers. This remarkable evolution creates unique opportunities for amateur astronomers to make significant contributions to a broadening of spectroscopic survey programs. Nowadays these activities are further supported by excellent freeware programs documented by detailed manuals, which are even supplemented with step-by-step tutorials, provided by numerous Internet sources. This evolution in spectroscopic hard- and software stimulates the rise of symbiotic cooperation within scientific projects between professional and amateur astronomers.

Meanwhile, spectroscopic investigations have even surpassed the importance of photometric surveys, which in earlier days formed our main contribution to astronomical science. This discipline has now more and more been taken over by powerful professional satellite programs. On the one hand this is of course a huge benefit for research but on the other the scientific relevance of this classical amateur activity is now sadly reduced. Nevertheless some niches still remain here for useful photometric surveys not least as a valuable supplement for ongoing spectroscopic campaigns.

This book is intended to support the pleasing upturn of spectroscopy and is mainly addressed to slightly advanced amateurs to enable the highly rewarding "upgrade" from merely photographer to real analyst of astronomical spectra. For this purpose the most important astrophysical applications are presented here, which are accessible to rather modest amateur equipment, allowing reproduction of many of the significant discoveries of modern astronomy. The *Spectral Atlas* [1] is supplemented and supported here by the theoretical basics, thus enabling a deeper understanding of the processes involved which are related to the various objects presented. In Section 15.1 a rewarding opportunity is presented for amateurs to contribute to highly motivating mixed professional/amateur projects.

A further focus here is on theoretical and practical aspects of the processing, calibration and normalization of spectral profiles and their most important analysis and measurement options. For a better understanding and an optimized application of our "work horses," the optical and physical basics about the most common types of spectrographs are provided. Special attention is devoted to the mechanical and optical aspects and their impacts on the specifications of the telescope and the recording CCD camera.

The first chapter is dedicated to the physical basics of spectroscopy. The main topics are the properties of photons, blackbody radiation, spectroscopic wavebands, Plank's energy equation and the formation and typology of astronomical spectra. Unfortunately, the quantum mechanical models required to explain the possible levels for the individual electron transitions which are directly responsible for the specific wavelengths of the generated spectral lines are highly complex and abstract. Within the scope of this book, the section concerning the Schrödinger equation is directed particularly to students and amateurs with a deeper interest, requiring advanced knowledge of mathematics and quantum mechanical terminology. However, for the practical work this section may be skipped without losing the thread of the story.

Many amateurs, who are principally interested in spectroscopy, erroneously believe that they would lack the necessary background to meet the theoretical requirements. However, with a few exceptions, the presented astrophysical applications only require basic understanding in physics and algebra as well as the ability to calculate simple formulas using a scientific pocket calculator. Nevertheless, it would be wrong to say that spectroscopy is suitable for everybody. What is required is a deeper interest in specific activities like processing and analyzing, combined with a basic affinity for physics and a great deal of scientific curiosity. In contrast to high-quality photos of deep sky objects, and although undoubtedly present, the tremendous aesthetic aspects of astronomical spectra are not that obvious and need some knowledge to be revealed. The necessary practical effort, however, compared to an ambitiously practiced astrophotography, is by far lower and the recording of the spectra considerably less sensitive in respect of the seeing quality.

On a first glance it may be surprising that even the necessary chemical knowledge remains pretty limited. However, in the hot stellar atmospheres and excited nebulae the individual elements can hardly undergo any form of chemical bonding. Only in the outermost layers of relatively "cool" stars some very simple, mostly just diatomic, molecules can survive. More complex chemical compounds are found only in the cooler and unionized outskirts of HII regions, within circumstellar envelopes and cocoons, which make up part of cold dust clouds in the interstellar space as well as in planetary atmospheres – a typical domain of radio astronomy, microwave spectroscopy or planetary space probes. Furthermore, the abundance of hydrogen and helium in the visible matter in the Universe is still about 99 percent. In stellar astronomy, all elements except hydrogen and helium are simplistically designated as "metals."

Finally, we hope that this book, combined with the *Spectral Atlas*, will be helpful to many amateurs worldwide. In this way we also hope to open the door of an exciting world – enabled by own means, to acquire an amazingly detailed insight in to the properties of astronomical objects which may be up to billions of light years away. Although – as actually generally accepted – only roughly 4 percent of the Universe is spectroscopically visible to us, it makes much worth to explore that part of the cosmos more in detail.

ACKNOWLEDGEMENTS

While working on the *Spectral Atlas* the necessity for a separate book soon became obvious. A book that would relieve the *Atlas* of theoretical and practical matters. This way the amateur astronomer has at their disposal a second book, which provides supplementary information and interesting astrophysical applications, customized to their specific needs. Used in parallel to the *Atlas* it fills the gap between theoretical knowledge and practical astronomical spectroscopy, simultaneously opening the way to a more professional approach, a *conditio sine qua non* for a successful participation in professional/amateur spectroscopy projects. For this multidisciplinary task we wish to thank all the people involved for their valuable advice and cooperation. The numerous emails between us and colleagues, amateurs, professionals, companies and friends have been a tremendous help for the fine-tuning of this work. Therefore the many people already involved and mentioned in this context with respect to the realization of the *Spectral Atlas*, deserve our acknowledgement here. However, we hope for understanding if not all are mentioned individually again here.

Special thanks go to the company representatives or manufacturers of astronomical spectrographs, being: Johannes and Thomas Baader (Baader Planetarium, Germany), Martin Huwiler (Eagleowloptics, Switzerland), Terry Platt (Starlight Xpress, UK), Daniel Sablowski (Astro Spectroscopy Instruments, Germany), Olivier Thizy (Shelyak, France), Mark Woodward and Ken Harrison (JTW Astronomy, the Netherlands) for their willingness to provide detailed technical information and illustrations on their spectrographs. We are also grateful for their continuous efforts to offer entry level and research grade spectrographs for personal, educational and scientific cooperative projects. Additionally a special recognition is given here to Martin Huwiler, who substantially contributed to several chapters, containing optical and/or practical aspects.

An especially high effort was required for the search for information about the calibration of the spectral flux density, this revealed a lack of appropriate publications, treating this topic not just fragmentary but rather comprehensively. To provide here a somewhat reasonable overview, tailored to the needs of the amateur, the information had tediously to be gathered from numerous sources. In addition further supplementary inputs had to be obtained, and many points to be clarified, by intense and sometimes even controversial discussions with many amateurs and even some professionals. Thus specific thanks deserve here all, having proved patience and contributed in any kind to this chapter.

Our deepest gratitude goes to our wives, children and other family members for their support, interest and patience during all phases before and under the editorial process. Finally of course we generally want to express our warmest thanks to everyone who contributed in any kind to this book.

CHAPTER

1

Physical Basics of Spectroscopy

1.1 Photons: Carriers of Information and Energy

1.1.1 Photons: Carriers of Information

Photons are extremely important for all astronomy because they carry information about the observed object to our recording devices. Some of them finally end up in the pixel field of our CCD camera. It is therefore worthwhile to take some time to consider this absolutely most important link in the transmission chain.

It was on the threshold of the twentieth century, when the discovery of the photon with its seemingly odd properties, caused tremendous "headaches" to the entire community of top physicists. This huge intellectual effort finally culminated in the development of quantum mechanics. The list of participants reads substantially like the *Who's Who* of physics at the beginning of the twentieth century: Werner Heisenberg,

Albert Einstein, Erwin Schrödinger, Max Born, Max Planck, Wolfgang Pauli, Niels Bohr, just to name a few. Quantum mechanics became, besides the theory of relativity, the second revolutionary theory of the twentieth century. For the rough understanding about the formation of the photons and finally of the spectra, the necessary knowledge may be reduced to some very basic key points of this theory.

1.1.2 The Wave–Particle Duality

Electromagnetic (EM) radiation exhibits both wave and particle nature. This principle applies to the entire spectrum. Starting with the long radio waves, it remains valid on the domains of infrared radiation, visible light, up to the extremely short-wave ultraviolet, X-rays and gamma rays (Figure 1.1). For our present technical applications, both properties are indispensable. For the entire telecommunications, radio, TV,

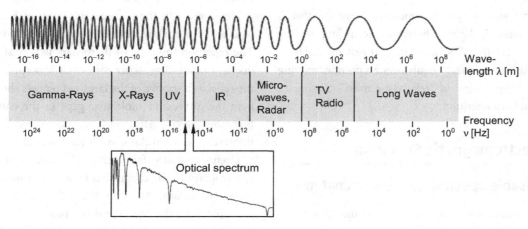

Figure 1.1 Electromagnetic spectrum

mobile telephony, as well as the radar and microwave grill it is the wave nature that is utilized. However, CCD photography, light metering on cameras, gas discharge lamps (e.g. for energy saving light bulbs and street lighting) and last but not least spectroscopy all require the particle nature to work.

1.1.3 The Quantization of Electromagnetic Radiation

It was one of the pioneering discoveries of quantum mechanics that electromagnetic radiation is not emitted continuously but rather quantized (or quasi "clocked"). Simplified explained a minimum "dose" of electromagnetic radiation is generated, called "photon," which belongs to the bosons within the "zoo" of the elementary particles. The term "photon" derives from the Greek "$\varphi\tilde{\omega}\varsigma$" or "phôs," which means light.

1.1.4 Photons: Carriers of Energy

Each photon has a specific frequency (or wavelength), which proportionally determines the transported energy – the higher the frequency, the higher the energy of the photon (see Section 1.3.3).

1.1.5 Other Properties of Photons

Without any external influences photons exhibit an infinitely long life expectancy. Their generation and "destruction" takes place in a variety of physical processes. Chiefly relevant for astronomical spectroscopy is the broadband generation of photons by hot stars and the narrow-band emission and absorption due to electron transitions between different atomic orbitals.

A photon carrying a specific amount of energy always moves with speed of light. Therefore, according to the special theory of relativity (STR), it has no rest mass. However, based on the mass–energy equivalence a photon, moving with light speed, has a kind of "relativistic mass" causing an actual measurable momentum.

1.2 The Electromagnetic Spectrum

1.2.1 The Usable Spectral Range for Amateurs

Professional astronomers nowadays study the objects in nearly the entire electromagnetic spectrum – including the domain of radio astronomy (see Figure 1.1).

Furthermore, space telescopes are used which are increasingly optimized for the infrared domain in order to record the extremely redshifted spectra of objects from the early days of the Universe. For the ground-based amateur, equipped with standard telescopes and spectrographs, only a modest fraction of this domain is reachable. In addition to the specific design features of the spectrograph, the usable range is limited here mainly by the spectral characteristics of the camera, including any filters. So, for example, the camera Atik 314L+ combined with the DADOS spectrograph, achieves useful results in the range of approximately λλ3800–8000, i.e. throughout the visible domain and the near infrared part of the spectrum. Here also the well known lines are located, such as the H-Balmer series and the Fraunhofer absorptions.

1.2.2 The Selection of the Spectral Range

For high-resolution spectra, the choice of the spectral range is normally determined by a specific monitoring project or an interest in particular lines. Perhaps also the emission lines of the calibration light source may be considered in the planning of the recorded section. For low-resolution, broadband overview spectra, mostly the range of the H-Balmer series is preferred. Hot O and B stars can be taken rather in the short-wave part, because their radiation intensity is very strong in the UV range. Only in special cases it makes sense here to include an area on the red side of Hα. Between approximately λλ6200 and 7700 [1] it is swamped with atmospheric related (telluric) H_2O and O_2 absorption bands. Apart from an undeniable aesthetic aspect they are interesting only for the atmospheric physicist. For astronomers, they are usually only a hindrance, unless the fine water vapor lines are used to calibrate the spectra! Anyway, there exist methods to extract these absorptions from the profile, for example, with the Vspec software or with the freeware program SpectroTools [73].

However, with the late spectral types of K and the entire M class [1] it makes sense to record the near infrared range. The radiation intensity of these stars is very strong here and displays interesting stellar molecular absorption bands. Also, the reflection spectra of the large gas planets show particularly here the impressive molecular gaps in the continuum. Further, the *Spectral Atlas* [1] is an aid to find the appropriate interesting spectral domain for all object classes.

Useful guidance for setting the wavelength range of the spectrograph is the calibration lamp spectrum or the daylight (solar) spectrum. At night the reflected solar spectrum is available from the Moon and the planets. A striking marker on the blue side of the spectrum is the impressive double line of the Fraunhofer H and K absorption.

Table 1.1 Terminology in the optical range (UBVRI λλ3300–10,000)

Center wavelength			
λ [μm]	λ [Å]	Astrophysical wavebands	Required instruments
0.35	3500	**U** band (UV)	
0.44	4400	**B** band (blue)	
0.55	5500	**V** band (green)	Most optical telescopes
0.65	6500	**R** band (red)	
0.80	8000	**I** band (infrared)	

Further in use is also the Z band, ~λλ8000–9000 and the Y band, ~λλ9500–11,000 (ASAHI Filters).

Table 1.2 Terminology in the infrared range (λλ10,000–2,000,000)

Center wavelength			
λ [μm]	λ [Å]	Astrophysical wavebands	Required instruments
1.25	12,500	**J** band	
1.65	16,500	**H** band	Most optical and dedicated infrared telescopes
2.20	22,000	**K** band	
3.45	34,500	**L** band	
4.7	47,000	**M** band	Some optical and dedicated infrared telescopes
10	100,000	**N** band	
20	200,000	**Q** band	
200	2,000,000	Submillimeter	Submillimeter telescopes

Source: Wikipedia page on infrared astronomy

1.2.3 Terminology of the Spectroscopic Wavebands

Terminology for wavebands is applied inconsistently in astrophysics [7] and depends strongly on the context. Furthermore many special fields of astronomy, various satellite projects etc. often use different definitions. Tables 1.1 and 1.2 give a summary according to [7] and Wikipedia (infrared astronomy). Indicated are either the center wavelength λ of the corresponding photometric band filters, or their approximate passband. The original passband of these broadband filters, applied for the photometric UBVRI system, have been defined by Johnson, Bessel and Cousins.

Table 1.3 Terminology predominantly applied by ground-based observatories

Far Ultraviolet (FUV)	λ <3000	Satellite based special telescopes
Near Ultraviolet (NUV)	λ 3000–3900	
Optical (VIS)	λ 3900–7000	
Near Infrared (NIR)	λ 6563 (Hα)– 10,000	
Infrared or Mid-Infrared	λ 10,000–40,000	(J, H, K, L band 1–4 μm)
Thermal Infrared	λ 40,000–200,000	(M, N, Q band 4–20 μm)
Submillimeter	λ >200,000 (200 μm)	

1.3 Wavelength and Energy

1.3.1 Preliminary Remarks

According to the recommendations of the International Astronomical Union (IAU) among others, the cgs system (centimeter, gram, second) along with the units [erg], angstrom [Å] and gauss [G] should no longer be applied. However, the angstrom for the wavelength is still in use (and not only in the amateur sector). Furthermore in astrophysical papers, for example, the surface gravity g and many other applications are still expressed in cgs units and the magnetic flux density in gauss [G].

1.3.2 Units for Energy and Wavelength Applied in Spectroscopy

It is still very common for many applications to use the erg rather than the joule [J] and $1\,erg = 10^{-7}$ J. Thus for the line flux the unit [erg s^{-1} cm^{-2}] is widely in use (see Section 9.1.3). For the extremely low energies of electron transitions instead of joule [J] almost always the unit electronvolt [eV] is applied, in accordance with the IAU recommendations and $1\,eV = 1.602 \times 10^{-19}$ J.

Furthermore, in the optical spectral domain the wavelengths are extremely small and in astronomy they are usually measured in angstroms [Å] or nanometers [nm]. One should be aware that 1 Å roughly corresponds to the diameter of an atom, including its electron shells! In the infrared range the micron [µm] is also in use: $1\,\text{Å} = 10^{-10}\,\text{m}$, $1\,\text{nm} = 10\,\text{Å}$, $1\,\text{µm} = 1000\,\text{nm} = 10{,}000\,\text{Å}$.

Sometimes, the wavelength λ is also expressed as "wavenumber k," which is the reciprocal value of λ. Here we mainly see k expressed in number of waves within $1\,\text{cm}$ [cm^{-1}], a further example of the cgs system has obviously still "survived."

$$k = \frac{1}{\lambda} = \frac{\nu}{c}\,[\text{cm}^{-1}] \qquad \{1.1\}$$

1.3.3 Planck's Energy Equation

For each individual line the corresponding photon energy E can be calculated. This is enabled by the simple equation of the German physicist and Nobel Prize winner Max Planck (1858–1947):

$$E = h\nu \qquad \{1.2\}$$

Here, E = photon energy in joule [J], h = Planck's constant (the quantum of action $6.626 \times 10^{-34}\,\text{J s}$, ν Greek "nu") = frequency [s^{-1}] of the photon of the spectral line.

The frequency ν is simply related to the wavelength λ[m] (c = speed of light $3 \times 10^{8}\,\text{m s}^{-1}$):

$$\nu = \frac{c}{\lambda} \qquad \{1.3\}$$

Inserting Equation {1.3} into {1.2} yields:

$$E = \frac{hc}{\lambda} \qquad \{1.4\}$$

The most important statement of Equations {1.2} and {1.4}: The energy of a photon E is proportional to its frequency ν and inversely proportional to the wavelength λ.

The following simple equations, suitable for pocket calculators, allow converting the wavelength λ [Å] into energy E [eV] and vice versa:

$$\lambda[\text{Å}] = \frac{12403}{E[\text{eV}]} \qquad \{1.5\}$$

$$E[\text{eV}] = \frac{12403}{\lambda[\text{Å}]} \qquad \{1.6\}$$

1.4 The Continuum and Blackbody Radiation

1.4.1 The Blackbody as a Physical Model for Stellar Radiation

Heating up an object raises its temperature, which determines the movements of the atoms inside the object. When the temperature becomes higher than the environment, the object starts to emit electromagnetic radiation. Objects around us, inclusive ourselves, also reflect radiation, which is noticeable in the visible or infrared wavelength range.

Stars, like our Sun behave differently in that they do not reflect radiation which falls on them, but rather absorb all incident radiation. Therefore, as an approximation, but with some exceptions, we can consider the Sun as a blackbody.

A perfect blackbody absorbs all radiation that falls on it, does not reflect any radiation nor is transparent to it. This means that blackbody radiation, i.e. the emitted radiation of a blackbody, is exclusively determined by its temperature. As a result of this approximation, measurement of the emitted radiation of a star lets us determine its temperature through the specific distribution of the blackbody radiation (see Figure 1.3).

The physical model of the blackbody applies well to the non-transparent part of the stellar photosphere. Other parts of the stellar atmosphere show a different behavior as the atoms present absorb photons at certain wavelengths. This represents a highly important source of information, included in the generated spectrum of the star. In contrast to the distribution of the blackbody radiation the analysis of the star's spectrum gives us insight into the chemical composition.

1.4.2 Planck's Radiation Law and Course of the Continuum Level

The curve in Figure 1.2, hereafter referred to as continuum level $I_C(\lambda)$, corresponds to the course of the spectral flux density (see Section 9.1.2), sometimes abbreviated as intensity I. It is plotted over the wavelength, which increases from left to right. In the original, undisturbed profile it represents very roughly the exclusively temperature dependent blackbody radiation characteristics $B_{T\text{eff}}(\lambda)$ of the star. As a fit to the continuum it is cleaned from any existing absorption or emission lines. The entire area between the horizontal wavelength axis and the continuum level $I_C(\lambda)$ is called "continuum." This blackbody model is quite useful for rather cool stars exhibiting a maximum radiation in the green to red part of the spectrum, but excluding the M class with the strong and very uneven TiO absorptions [1]. For hot stars, as shown in Figure 1.2, radiating mainly in the blue up to the UV range, it gets significantly overprinted by the Balmer jump of the hydrogen series (see also Figures 2.4 and 2.5).

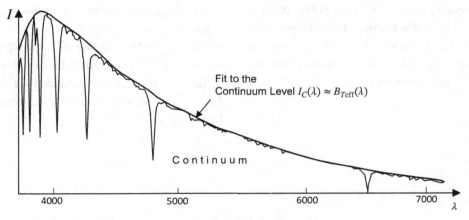

Figure 1.2 Continuum, continuum level and approximate blackbody radiation $B_{T\text{eff}}(\lambda)$

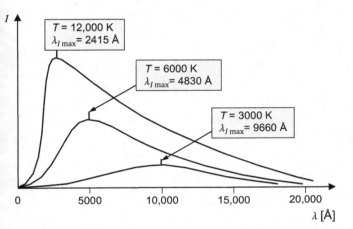

Figure 1.3 Planck radiation and Wien's displacement law

1.4.3 Wien's Displacement Law

Figure 1.3 shows the theoretical blackbody radiation distribution of different stars, exhibiting bell-shaped curves $B_{T\text{eff}}(\lambda)$. With increasing temperature their peak intensities get shifted to shorter wavelength, and respectively to higher frequency (Planck's radiation law). With Wien's displacement law and the given wavelength [Å] of the maximum radiation intensity $\lambda_{I\max}$ [Å] it is theoretically possible to roughly estimate the "effective temperature," T_{eff} [K], of a star (Wilhelm Wien (1864–1928) was a German physicist and Nobel Prize winner):

$$\lambda_{I\max}\,[\text{Å}] \approx \frac{28{,}978{,}200}{T_{\text{eff}}} \qquad \{1.7\}$$

$$T_{\text{eff}}\,[\text{K}] \approx \frac{28{,}978{,}200}{\lambda_{I\max}} \qquad \{1.8\}$$

Examples of the wavelengths of maximum blackbody radiation:

Alnitak	$T_{\text{eff}} \approx 25000$ K	$\lambda_{I\max} \approx 1160$ Å (ultraviolet range)
Sun	$T_{\text{eff}} \approx 5800$ K	$\lambda_{I\max} \approx 4996$ Å (green range)
Betelgeuse	$T_{\text{eff}} \approx 3450$ K	$\lambda_{I\max} \approx 8400$ Å (infrared range)

1.4.4 Effective Temperature T_{eff} and Stefan–Boltzmann Law

The effective temperature, T_{eff}, is the temperature required for a blackbody radiator with the same size of the star, in order to generate the identical bolometric energy flow F_{Bol} [erg cm^{-2} s^{-1}]. However, F_{Bol} is not limited to the visual spectral range but the radiation energy per unit time and unit area corresponds to the total area under the Planck radiation curve. It is obtained by integration of the spectral flux density $I(\lambda)$ over the entire electromagnetic spectrum from $\lambda = 0$ to infinity (∞):

$$F_{\text{Bol}} = \int_{0}^{\infty} I(\lambda)\,d\lambda \qquad \{1.9\}$$

Further, according to the law of Stefan–Boltzmann, F_{Bol} is proportional to the fourth power of the effective temperature T_{eff}

$$F_{\text{Bol}} = \sigma\,T_{\text{eff}}^4 \qquad \{1.10\}$$

where σ is the Stefan–Boltzmann constant: 5.67×10^{-5} erg s^{-1} cm^{-2} K^{-4}. However, it must never be confused with the Boltzmann constant k_{B}, applied in statistical mechanics (see Section 14.2.3). The Stefan–Boltzmann law is named after the Austrian physicist and mathematician Josef Stefan (1835–1893) together with his compatriot, the physicist and philosopher Ludwig Boltzmann (1844–1906). If F_{Bol} of a star with the radius R is multiplied with its supposed spherical surface $4\pi R^2$, it yields the entire bolometric luminosity L in [erg s^{-1}], under the assumption that the emission takes place evenly distributed:

$$L = 4\pi R^2 \sigma T_{eff}^4 \qquad \{1.11\}$$

For example the bolometric luminosity of the Sun with a temperature of ~5780 K and a radius of ~695,800 km, yields $L \approx 3.846 \times 10^{33}$ erg s^{-1}, corresponding to 3.846×10^{26} W. The solar flux density finally reaching the surface of the Earth is defined as the solar constant $E_0 = 1368$ W m^{-2} (from the World Meteorological Organization, WMO).

2 Electron Transitions and Formation of the Spectra

2.1 Simple Textbook Example: The Hydrogen Atom

The orbital energy diagram in Figure 2.1 shows the simplest possible example, the hydrogen atom, the fixed grid of the energy levels "n," which a single electron can occupy in its orbit around the atomic nucleus. They are identical with the shells of the famous Bohr atomic model and are designated by the so called principal quantum numbers. Which level the electron currently occupies depends on its state of excitation. A stop between the orbits is extremely unlikely. The lowest level is $n = 1$. It is closest to the nucleus and also called the ground state. With increasing n number (in Figure 2.1 from bottom to top):

- increases the distance to the nucleus
- increases the total energy difference, in relation to $n = 1$
- the distances between the levels and thus the required energy values to reach the next higher level, are getting smaller and smaller, and finally tend to zero on the level $E = 0$ eV (or $n = \infty$).

The energy level E on the level $n = \infty$ is physically defined as $E = 0$ eV [13] and also called ionization limit. The level number $n = \infty$ is considered as "theoretical," because a limited number of ~200 is estimated, which a hydrogen atom in the interstellar space can really occupy [15]. By definition, with decreasing n number the energy becomes increasingly negative. Above $E = 0$ eV, i.e. outside of the atom, it becomes positive.

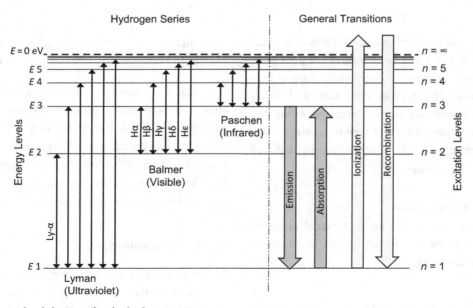

Figure 2.1 Orbital energy level diagram for the hydrogen series

2.2 Transition Types and Probabilities

Movements of electrons between the discrete quantized energy levels are called transitions. The nature of a generated spectrum is defined by the different types of transitions. Apart from the population on each energy level the intensity of the spectral lines is also influenced by the probability for the transitions to occur.

2.2.1 Absorption

A first type of transition is absorption and describes the movement of electrons from a lower energy level L to a higher level H by absorbing a photon. *Absorption* occurs only when the atom meets a photon whose energy matches exactly to a level difference by which the electron is then briefly raised at the higher level (resonance absorption). This type is called "bound–bound" transition, because the electron remains within the shell structure of the atom. The rate of transitions for absorption is given by Equation {2.1}:

$$R_{Abs} = B_{L \to H} N_L \rho, \qquad \{2.1\}$$

where R_{Abs} is the transition rate for absorption which is the number of transitions per unit of time, N_L represents the population number of the lower energy level, ρ is the energy density, $B_{L \to H}$ is the probability of transition from level L to H, also called the Einstein B coefficient. It is expressed in units of $[\mathrm{cm^3\, erg^{-1}\, s^{-1}\, Hz^{-1}}]$ or $[\mathrm{m^3\, J^{-1}\, s^{-2}}]$. The rate of transition by absorption is according to Equation {2.1} proportional to the energy density.

2.2.2 Emission

A second type of transition is emission, which occurs when the electron falls back to a lower level, emitting a photon in any random direction, but whose energy exactly corresponds to the level difference. This type is also called a "bound–bound" transition, because the electron remains within the shell structure of the atom. With this type of transition two different mechanisms can occur, spontaneous emission and induced or stimulated emission.

2.2.2.1 Spontaneous emission

Spontaneous emission occurs without an external radiation field. In that case the rate of transition for spontaneous emission is given by Equation {2.2}:

$$R_{Em_{sp}} = A_{H \to L} N_H \qquad \{2.2\}$$

Where $A_{H \to L}$ is the transition probability for emission also called the Einstein A coefficient and N_H the population number

of the higher energy level. It is independent of the energy density and is expressed as units of $[\mathrm{s^{-1}}]$. The inverse of A is defined as the lifetime, τ, of the state. A relatively high transition probability corresponds to a relatively short lifetime, and vice versa.

2.2.2.2 Induced or stimulated emission

Emission in the presence of an external radiation field is called *induced* or *stimulated*. Here, an incoming photon causes the emission, hence the name stimulated. This results in two outgoing photons which follow the same direction. Here, the rate of transitions by stimulated emission is given by Equation {2.3}:

$$R_{Em_{st}} = B_{H \to L} N_H \rho \qquad \{2.3\}$$

where $B_{H \to L}$ is the probability of transition from level H to L. In contrast to the spontaneous emission it is again dependent on the energy density! Therefore the units for probability of stimulated emission are comparable with those for absorption [102].

2.2.3 Ionization

Ionization of an atom occurs if the excitation energy is high enough to lift the electron above the level $E = 0\ \mathrm{eV}$ and it is forced to escape into space. This type is therefore called "bound–free" transition, because the electron leaves the shell structure of the atom. This event can be triggered by:

- an encounter with a high-energy photon: photoionization
- heating: thermal ionization
- a collision with external electrons or ions: collision ionization

Similar as for non-ionized atoms or molecules the different energy levels of ionized atoms and molecules create the possibility for different transitions, each characterized by their corresponding probabilities. A typical example is double oxygenated oxygen O^{++} known for its important [OIII] transitions used as one of the temperature diagnostics in gaseous nebulae and active galactic nuclei (AGN) [1].

2.2.4 Recombination

Recombination occurs if an ionized atom recaptures a free electron from the surrounding space and becomes "neutral" again. This type is called a "free–bound" transition because the electron enters the shell structure of the atom.

2.2.5 Electrons in the Free–Free Mode

Free electrons in the so called "free–free" mode may also absorb photons, gaining energy in this way. In the reverse case they lose energy by emitting a photon. The case of emission is here called "bremsstrahlung," typically caused by, for example, deceleration by the interaction with magnetic or electric fields. This term originates from the German words for "bremsen" (to brake) and "strahlung" (radiation).

2.3 Definitions and Notation

2.3.1 Ionization Stage versus Degree of Ionization

The term "ionization stage" refers here to the number of electrons, which an ionized atom has lost to space (Si IV, Fe II, H II, etc.). This must not be confused with the term "degree of ionization" in plasma physics. For a certain element in a gas it defines the ratio of the particle number densities of ionized to unionized atoms, as a function of the temperature, ionization energy and further parameters. This ratio is determined in astrophysics with the Saha equation derived by the Bengali astrophysicist Meghnad Saha (1893–1956).

2.3.2 Astrophysical Notation for the Ionization Stage

In astrophysics the chemical form of notation is not applied but instead another somewhat "misleading" version. Chemists denote neutral hydrogen by H, and ionized with H^+, which is clear and unambiguous. But astrophysicists denote neutral hydrogen with a Roman numeral as H I and the ionized with H II. The doubly ionized calcium is referred to by chemists as Ca^{++} or Ca^{2+}, for astrophysicists this corresponds to Ca III. For the 13 times ionized Fe XIV, typically occurring in spectra of nova outbursts, this corresponds to Fe^{13+}. In any case a high ionization stage of atoms always means that very high temperatures must be involved in the process.

2.4 The Hydrogen Spectral Series

A group of electron transitions between a fixed energy level and all higher levels is called a "transition series." In the middle of Figure 2.1 the Balmer series is displayed as a group of arrows. For amateurs it is of particular importance, because exclusively these hydrogen lines are within the visible range of the spectrum. They include all the electron transitions, which in the case of absorption, starting upwards from the second-lowest energy level $n = 2$. In the case of emission they end here by "dropping" down from an upper level. The Balmer series was discovered and described by the Swiss mathematician (and architect!) Johann Jakob Balmer. The lines of the adjacent Paschen series lie in the infrared and those of the Lyman series in the ultraviolet range.

The easiest way to display the Balmer series is to record a stellar spectrum of the spectral class A, for example Sirius (A1) or Vega (A0). Their effective temperature of about 10,000 K is best suited to generate impressively strong H-Balmer absorptions. At this temperature the share of electrons, which – just by thermal excitation – already occupies the base level $n = 2$, reaches a maximum. At still higher temperatures, this share decreases, because it gets shifted to even higher levels, for example, of the Paschen series. Finally the electrons are completely released, which results in the ionization of the H atoms.

2.4.1 The Photon Energy at the Wavelengths of the H-Balmer Series

Figure 2.2 shows schematically a Sirius-like spectrum with the first six H-Balmer lines, supplemented with a scheme of the relevant electron transitions. The absorption lines are consecutively labeled with lowercase Greek letters, starting with Hα in the red region of the spectrum, which is generated by the lowest transition $n = 2$ to $n = 3$. From Hε upwards often the respective level number n is used, for example H$\zeta \triangleq$ H8. Impressively to see here, how in the blue area the gaps between the absorptions get closer and closer – a direct consequence of the decreasing amount of energy, which is required to reach the next higher level. Perhaps this may be the most aesthetical phenomenon the spectroscopy has optically to offer!

Figure 2.2 also shows the photon energy E [eV] for the wavelengths of some hydrogen lines, calculated with Equation {1.6}. This is the way spectra in the UV domain are often calibrated in professional publications. On the left side the H-Balmer series is limited by the so-called "Balmer edge" or "Balmer jump," also referred to as "BJ" or "Balmer discontinuity." It ends at ~λ3646 and the continuum suffers a dramatic drop in intensity. This is caused by the huge phalanx of highly concentrated and increasingly closer following absorption lines, acting here as a barrier to photons of this wavelength domain. For the BJ this means photons are lost to ionization on the hydrogen $n = 2$ level.

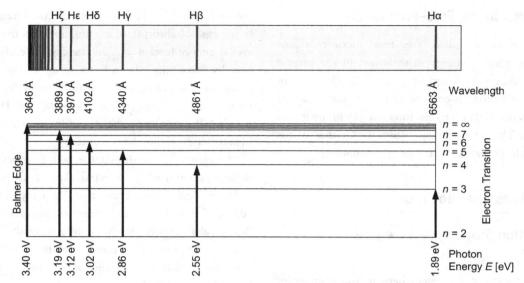

Figure 2.2 Sirius-like spectrum with H-Balmer lines, the corresponding electron transitions and photon energies

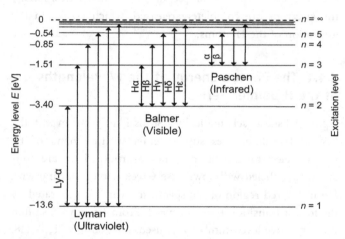

Figure 2.3 Quantified orbital energy level diagram

2.4.2 The Photon Energy of Electron Transitions

The photon energy E of a spectral line also corresponds to the energy difference ΔE between the initial and excited level of the causal electron transition. Therefore it also fits to the arrow lengths in the orbital energy level diagrams, for example in Figure 2.2. Here, 2.55 eV corresponds to the transition $n = 2$ to $n = 4$, or Hβ. This relationship enables us to calculate the energy values for the individual levels in the diagram of Figure 2.3.

2.4.3 Quantified Orbital Energy Level Diagram of the H-Balmer Series

By definition the energy at level $n = \infty$ is set to $E = 0$. Thus the corresponding value of 3.40 eV is shifted, with negative sign, down to the initial level of the Balmer series $n = 2$. Here it represents the required, additional energy to ionize

a hydrogen atom, starting from this level. From the lowest level $n = 1$ with 13.6 eV significantly more energy would be required. For the calculation of the individual excited levels the photon energy of the corresponding H-Balmer line must now be added to –3.40 eV. For example, for $n = 3$ this yields –3.40+1.89 = –1.51 eV, corresponding to the excited level of Hα. In the next chapter a general equation {3.18} will be derived to calculate the individual excited energy levels for the hydrogen atom.

2.4.4 The Lyman Limit of Hydrogen

According to Figure 2.3, the energy for the lowest ground state of the Lyman series is $E = -13.6$ eV. Converted with Equation {1.5}, it yields the well-known Lyman limit or Lyman edge in the UV range with wavelength λ912. It is functionally similar to the "Balmer jump" of the Balmer series (Figure 2.2). This value is very important for astrophysics because it defines the minimum required energy to ionize the hydrogen atom from its ground state $n = 1$. This level is only achievable by very hot stars of the O and early B class, with a temperature of $\gtrsim 25,000$ K [11]. The very high UV radiation of such stars ionizes first the surrounding hydrogen clouds which are shining due to emitted photons by the subsequent recombination. A textbook example is the H II region M42, in the Orion Nebula [1].

2.4.5 The Balmer and Rydberg Equations

Based on the four hydrogen lines Hα to Hδ visible in the optical domain, Johann Jakob Balmer developed empirically a simple mathematical law, enabling the rough wavelength calculation of the H-Balmer series with the corresponding level number n (e.g. Hα: $n = 3$). The formula further shows how this series, by increasing n values, converges to the wavelength of the Balmer edge A $\approx \lambda 3646$:

$$\lambda = A\left(\frac{n^2}{n^2 - 4}\right) \qquad \{2.4\}$$

The Swedish physicist Johannes Rydberg (1854–1919) generalized the Balmer formula, so that the wavelength of the other hydrogen series can be calculated:

$$\frac{1}{\lambda} = R\left(\frac{1}{n_1^2} - \frac{1}{n_2^2}\right) \qquad \{2.5\}$$

In its basic form it provides the reciprocal of the wavelength $[\lambda^{-1}]$ or wavenumber, as a function of:

- n_1, the base level of the electron transition and
- n_2, the level number of the excited state.

$R \approx 0.00109737315685$ Å^{-1} is the Rydberg constant for hydrogen. Written in this form it yields in Equation {2.5} the reciprocal of the wavelength directly in [Å^{-1}]. In 1888, using this formula, the existence of the other hydrogen series (e.g. Paschen, Lyman) could be predicted, although they were not yet experimentally detectable. Modified, it can also be applied to some other elements.

2.4.6 The Balmer, Paschen and Bracket Continua

Hot stars, with main radiation intensity in the UV region, show toward shorter wavelengths a steeply rising continuum

level. As shown earlier, this tendency is abruptly stopped by the Balmer jump at $\lambda 3646$. After a dramatic drop in intensity, the continuum rises again in the UV range, until it reaches the final ionization limit of hydrogen (also called the Lyman limit) at $\lambda 912$. Figure 2.4 shows the Balmer jump at a synthetic A0 I profile from the Vspec database.

A similar process occurs near the border to the infrared region at 8207 Å, the so-called Paschen jump (see Figure 2.5). Somewhat confusing is the designation of the intermediate continuum sections which always bear the name of the preceding jumps or series. The Balmer series n_2 to n_∞ is therefore located within the so-called Paschen continuum. On the shortwave (UV) side of the Balmer jump, follows the Balmer continuum with the Lyman series n_1 to n_∞. In the infrared region, the Paschen series, n_3 to n_∞, is located within the Bracket continuum. The dramatic influence of the hydrogen absorption on the continuum can be explained by the extremely high abundance of this element in most stellar photospheres.

2.5 Formation and Typology of Astronomical Spectra

2.5.1 Overview

Spectral lines are generated by the interaction between photons and atoms or molecules. Electron transitions are responsible for the characteristic narrow-band emissions or absorptions between different atomic orbitals. According to Kirchhoff's laws there are three basic types of spectra:

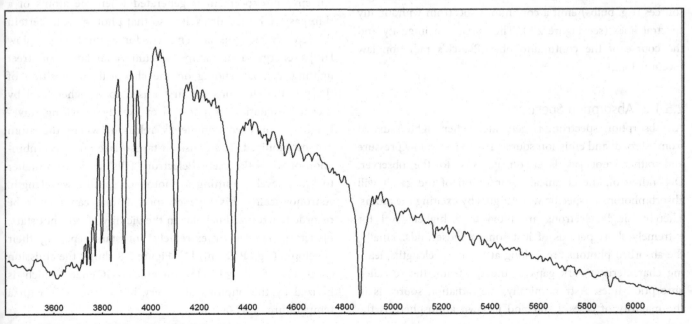

Figure 2.4 Synthetic A0 I profile with Balmer jump

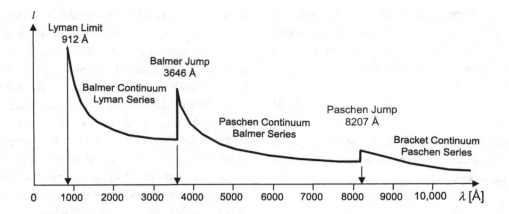

Figure 2.5 Balmer, Paschen and Bracket continua (schematically)

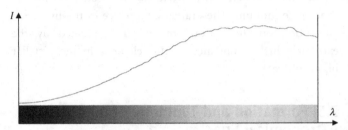

Figure 2.6 Continuous spectrum generated by halogen light bulb

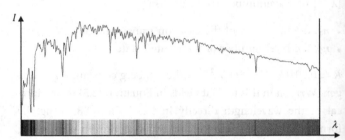

Figure 2.7 Photospheric absorption spectrum generated by the Sun (source: Database Vspec)

continuous, absorption and emission. These are summarized in the following subsections.

2.5.1.1 Continuous Spectrum

Similar to a blackbody radiator, hot incandescent light sources (e.g. bulbs) emit a continuous spectrum without any spectral lines (see Figure 2.6). The maximum intensity and the course of the continuum obey Planck's radiation law (Section 1.4.2)

2.5.1.2 Absorption Spectrum

An absorption spectrum is generated when light, radiated from a broadband emission source, has to pass a low pressure and rather cool gas layer on its way to the observer. Depending on the chemical composition of the gas it will absorb photons of specific wavelengths by exciting the atoms. Thereby single electrons are lifted to a higher level for extremely short periods of just some nanoseconds. Finally, the absorbed photons are lacking at these wavelengths, leaving characteristic dark gaps in the spectrum, the so-called absorption lines. Astronomically, the radiation source is in the majority of cases a star and its own atmosphere is the

comparatively "cooler" gas layer to be traversed. Figure 2.7 shows absorption lines in the solar spectrum.

2.5.1.3 Emission Spectrum

An emission spectrum is generated when the atoms of a thin gas are heated or excited so that photons with certain discrete wavelengths are emitted, for example neon glow lamps, energy saving lamps, sodium vapor lamps of street lighting, etc. Depending on the chemical composition of the gas, the electrons are first raised to a higher level by thermal excitation or photons of exactly matching wavelengths – or even completely released, where the atom becomes ionized. The emission takes place after recombination or when the excited electron "falls" back from higher to lower levels, emitting a photon of specific wavelength. Astronomically, this type of spectral line can mostly be recorded by ionized nebulae in the vicinity of very hot stars, planetary nebulae or extremely hot stars repelling their envelopes (e.g., P Cygni) [1]. Figure 2.8 shows the emission spectrum of the planetary nebula NGC 6826, which is ionized by the photons of a very hot central star with a temperature of 47,000 K.

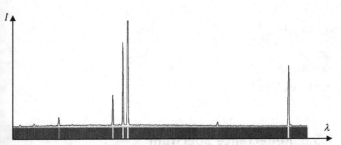

Figure 2.8 Emission spectrum generated by planetary nebula NGC 6826

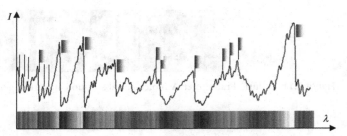

Figure 2.10 Absorption spectrum generated by the red giant α Her, spectral class M5Ib-II

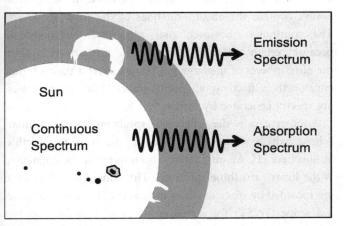

Figure 2.9 The three basic types of spectra, generated by the Sun.

2.5.2 The Three Basic Types of Spectra in the Context of the Sun

Figure 2.9 schematically shows the generation of the three basic types of spectra in the context of the Sun. The star itself generates a continuous spectrum. Observing the disk of the star, we see an absorption spectrum, generated by the light, penetrating on our line of sight its own slightly cooler atmosphere. Observing the stellar atmosphere, somewhat offside the apparent disk, we see emission lines, generated by the excited plasma.

2.5.3 Molecular Absorption Band Spectrum

Molecular absorption bands are a special case among the absorption spectra, generated by highly complex rotational and vibrational processes, excited by heated molecules. This takes place e.g. in the relatively cool atmospheres of red giants. The spectrum in Figure 2.10 was generated by the late classified Ras Algethi (α Her) spectral class M5Ib-II [1]. At this resolution it shows only a few discrete lines. The majority

are dominated by absorption bands, which are mainly caused by titanium monoxide (TiO) and to a lesser extent by magnesium hydride (MgH). In this case, these asymmetric absorptions reach the greatest intensity on the left, the short-wave band. Towards the right their intensity slowly weakens. The wavelength of absorption bands always refers to the point of greatest intensity the so called "band heads," also defined as "most distinct edge." As a result of this characteristic, sawtooth-shaped curve, these edges are much more diffuse than discrete absorption lines, which in theory show approximately a Gaussian bell shape. Depending on the source the indicated wavelength for the individual bands may sometimes vary, up to 2 Å! These absorption bands are therefore totally inappropriate for a precise calibration of the spectrum, based on known lines. Only the calibration with a light source can meet reasonable standards here.

Even several of the prominent Fraunhofer lines in the solar spectrum are mainly caused by telluric molecular absorption. Figure 2.11 shows the highly resolved O_2 absorption band spectrum of the Fraunhofer A line [1].

2.5.4 Molecular Absorption Band with Inversely Running Intensity Gradient

Figure 2.12 shows C_2 carbon molecular absorption bands in the blue–green region of the spectrum of the carbon star Z Piscium [1]. Generally for some carbon molecules (e.g. CO, C_2), the intensity gradient of the absorption bands runs in the opposite direction as with titanium monoxide (TiO) or O_2. By the middle of the nineteenth century this phenomenon had been recognized by Father Angelo Secchi. For this type of spectra, he introduced the "spectral type IV" [1].

2.5.5 Composite or Integrated Spectrum

Superimposed spectra of several light sources are called "composite" or sometimes also "integrated spectra." The term

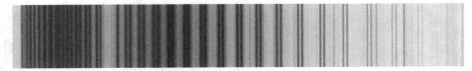

Figure 2.11 Telluric Fraunhofer A molecular absorption spectrum generated by sunlight [1]

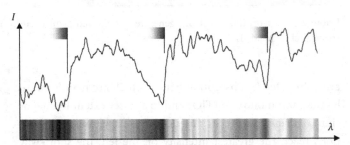

Figure 2.12 Cut out of the absorption spectrum generated by the carbon star Z Psc, spectral class C7,3 [1]

"composite" was coined in 1891 by Pickering for composite spectra in binary systems. Today it is often used for integrated spectra of stellar clusters, galaxies and quasars, which consist of from hundreds of thousands to several hundred billions of superposed spectra [1].

2.5.6 Reflectance Spectrum

The objects in our solar system are not self-luminous, but just visible due to reflected sunlight. Therefore, these spectra always contain the absorption lines of the solar spectrum. The continuum, however, may sometimes be modified, because certain molecules, for example methane (CH_4), in the atmospheres of the large gas planets exert a strong influence on the reflectivity at specific wavelengths. Special cases are spectra generated by comets [1].

Another case is the reflectance spectrum of a total lunar eclipse, which shows clearly the signatures of the Earth's atmosphere [1]. An alternative, also interesting for amateurs, is the lunar earthshine spectrum. The same types of spectra are recorded by space satellites such as the James Webb Space Telescope (JWST) for screening of exoplanets for possible bio-signatures.

CHAPTER

3 Quantum Mechanical Aspects of Spectroscopy

3.1 Quantum Mechanical View of Transition Probabilities

The relation between transition probabilities and quantum mechanics can be demonstrated by Equations {3.1} and {3.2}. The electron-wave interaction is considered as a dipole and the transition is interpreted in terms of wave functions according to Equation {3.7}. The transition probabilities, presented in Table 3.1, can then be expressed in terms of the dipole moment [103]:

$$B_{H \to L} = \frac{4\pi^3}{3\varepsilon_0 h^2} |p|^2 \qquad \{3.1\}$$

$$A_{H \to L} = \frac{2\omega^3_{HL}}{3\varepsilon_0 c^3 h} |p|^2 \qquad \{3.2\}$$

where $|p|$ is the transition dipole moment and ω_{HL} is the angular frequency for the transition between the energy states H and L. From Equations {3.1} and {3.2} some important conclusions can be made:

- Transition probabilities are proportional to the *absolute square of the dipole moment matrix element*, which influences the intensity of the spectral line corresponding to this transition.

- If one of the elements of the dipole moment matrix is zero the corresponding transition probability for that transition will also be zero.

- The transition probability for spontaneous emission is proportional to the *cube of the frequency*. This implies that this

Table 3.1 Transition types and probabilities

Transition Type	Transition Movements electron/photon	Symbol	Units
Spontaneous Emission		A	$[s^{-1}]$
Stimulated Emission		$B_{H \to L}$	$[cm^3\ erg^{-1}\ s^{-1}\ Hz^{-1}]$ or $[m^3\ J^{-1}\ s^{-2}]$
Absorption		$B_{L \to H}$	$[cm^3\ erg^{-1}\ s^{-1}\ Hz^{-1}]$ or $[m^3\ J^{-1}\ s^{-2}]$

type of emission has much greater probability to occur within the *shorter* wavelength ranges rather than the longer ones.

- Transition probabilities are typical properties of atoms, molecules and corresponding ionized forms independent from whether thermodynamic equilibrium is reached or not.
- The ratio of similar transition probabilities, for example, spontaneous emission can be used to calculate the expected intensity ratio of the corresponding spectral lines.

Databases of transition probabilities can be consulted on the website of the National Institute of Standards and Technology NIST [27].

3.2 The Energy Levels of Multi-Electron Atoms

For all other atoms, for instance with more than one electron, the description of line formation can be highly complex. All atoms have shell systems with different energy levels and are therefore distinguished by other wavelengths of the spectral lines. Another factor is the number of valence electrons on the outer shell, or how many of the inner levels are already fully occupied. Furthermore the main levels are subdivided in a large number of so-called "degenerate sublevels" which come into play for "housing" all the additional electrons. Each of those levels which have other implications in respect of quantum mechanics are presented in the next section. The energy differences between such sublevels must logically be very low. This explains why spectral lines of metals often appear in dense groups, a few with distances in between of even less than 1 Å! Typical examples in the solar spectrum are the sodium lines at $\lambda\lambda5896$ and $\lambda5890$ as well as the famous magnesium triplet at $\lambda\lambda5184$, 5173 and 5169. At least for amateur applications the hydrogen sublevels play no practical role, because they are "degenerate."

3.2.1 The Quantum Numbers *n, l, m, s* and Parity Operator P

The solution to the radial part of the Schrödinger equation gives the principal quantum number *n*, which correlates to the different energy levels according to Equation {3.18}. This system has been already presented in Section 2.1 which mainly defines the distance of the electrons orbiting the nucleus. Apart from the radial part of the Schrödinger equation here the angular part comes in to play. The solution to this provides for each principal quantum number *n* three

other, *hierarchically subordinated* levels of quantum numbers, whereas the orbitals around the nucleus may be considered as spherical of different size:

- azimuthal – *l*
- magnetic – *m*
- spin – *s*

With hydrogen the sublevels *l*, *m* and *s* are "degenerate."

3.2.1.1 Level of the Azimuthal Quantum Number *l*

Depending on the number of *n*, the azimuthal quantum number *l* takes values from 0 to *n* – 1. The first four individual *l* numbers are additionally designated with the following letters (see also Figure 3.7):

- *s* orbital for *l* = 0 (spherical shape with a maximum of 2 electrons)
- *p* orbital for *l* = 1 (dumbbell shape with a maximum of 6 electrons)
- *d* orbital for *l* = 2 (clover shape with a maximum of 10 electrons)
- *f* orbital for *l* = 3 (flower shape with a maximum of 14 electrons)

The electron configuration on the *l* level represents the orbitals of the atoms in different states by this so called "*spdf*" notation.

3.2.1.2 Level of the Magnetic Quantum Number *m*

The following sublevel of the magnetic quantum numbers *m* concerns the electron's orbital orientation in space exclusively when exposed to a magnetic field. Normally all *m* sublevels have the same energy and cannot be distinguished from each other, but all are nevertheless able to accommodate electrons. The magnetic sensitivity comes in to play only if an external magnetic field is acting, causing the Zeeman effect, which results in a split of spectral lines as outlined in Section 13.2.2. In any case there exists here a sublevel designated with *m* = 0. Further *m* levels are identified, for example for *l* = 2, with +1, –1, +2, –2. These values for *m* always range from –*l* to +*l*.

3.2.1.3 Level of the Spin Quantum Number *s*

The following final sublevel with the quantum number *s* describes the spin of the electron, which can be clockwise or counterclockwise. The two possible values of the spin quantum number are +1/2 and –1/2. For *l* = 0 the spins are degenerate and have equal energies. Figure 3.1 shows the hierarchical structure of the different quantum numbers for *n* = 4.

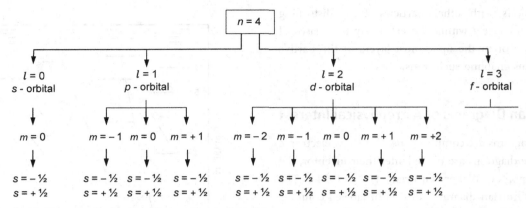

Figure 3.1 Hierarchical structure of orbitals and quantum numbers for $n = 4$

3.2.1.4 Pauli's Exclusion Principle

It is important to know that, considered at a certain time, each electron has its unique combination of quantum numbers, this is known as Pauli's exclusion principle. It was named after its discoverer, the Austrian born, Swiss physicist Wolfgang Pauli (1900–1958).

3.2.1.5 Hund's Rule

The different energy levels or orbitals within atoms or molecules are filled by electrons according the rule of maximal multiplicity, which is known as Hund's rule, named after the German physicist Friedrich Hund (1896–1997). Important to note here: Hund's rule is only applicable for atoms and molecules in the ground state.

3.2.1.6 Parity Operator P

How the spatial orientation of the wave function (see Section 3.3) behaves is analyzed by the parity operator P. Parity can be even (+) or odd (–) as a consequence of the inversion operation by P. The parity of a certain configuration can be calculated by summing all angular momentum quantum numbers of the active electrons, contributing to the concerned configuration.

3.2.2 Dipole Transitions, Laporte Rule and Forbidden Transitions

As the electrons, here considered as negatively charged particles, move around the positively charged nucleus, they cause changes in electric dipole moments as a consequence of interaction with an electromagnetic wave. A dipole is generated in atoms and molecules as a result of the uneven distribution of the electrons around the nuclei. When an electron makes a transition between different energy levels, the measured perturbation created in the electromagnetic field is given by the transition dipole moment. Transitions generated this way are called electric dipole transitions or simply dipole transitions, which appear as strong transitions. In spectroscopy they are the most commonly used transitions. Much weaker transitions are caused by magnetic dipoles or electric quadrupoles. For all kinds of transitions "selection rules" predict which transition is allowed or forbidden. If the transition dipole moment for a given transition is zero, it becomes a forbidden transition, to be interpreted as an extremely low probability for that transition to occur. If its value differs from zero the transition is allowed. In the case of dipole transitions they are only allowed if there is a change in parity. In spectroscopy this is called the Laporte rule, named after the German born, American physicist Otto Laporte (1902–1971). Textbook examples of Laporte's forbidden transitions are those of doubly ionized oxygen O^{2+} indicated as [O III] at $\lambda\lambda 4363, 4959$ and 5007 and the forbidden airglow lines [O I] at $\lambda\lambda$ 5577 and 6300. Excited by the solar wind they are involved in the formation of the greenish and reddish components of the polar lights in the extremely thin uppermost layers of the Earth's atmosphere. Forbidden lines are denoted within square brackets, for example [O III], [N II], [Fe XIV].

The initial levels of forbidden transitions are called "metastable," because they are highly sensitive to impacts and an electron must remain here for a quite long time (several seconds up to minutes) until it performs the forbidden "jump." These circumstances drastically increase the likelihood that this state is destroyed before the forbidden transition happens and due to an impact just an ordinary, allowed transition occurs instead.

Highly important for the interpretation of spectra is the simple knowledge that in dense gases, like near the Earth's surface or in stellar atmospheres, such forbidden transitions are extremely unlikely, because they are prevented by frequent collisions with other particles. Such disturbing effects occur very rarely within the extremely thin gases of interstellar space or in the uppermost layers of the Earth's atmosphere, thus enabling such transitions.

3.2.3 Grotrian Diagrams of Astrophysical Interest

Energy level diagrams are commonly used to show electronic transitions. The diagrams are named after their inventor, the German astronomer Walter Grotrian (1890–1954). The structure of a Grotrian diagram consists of labeled columns with horizontal lines representing the different quantum energy levels along a vertical axis scaled in electronvolts. Transitions between the different energy levels are indicated by perpendicular or slanted lines going upwards or downwards according to the absorption or emission process. The thickness of the lines represents the strength of the transition and the forbidden transitions are represented by dashed lines. The notation above the columns indicates the type of orbital and all lines are labeled according to their corresponding wavelengths. Note here that in case of multi-electron systems total quantum number orbital states are represented as capital letters: S, P, D, F, in contrast to the one-electron systems notation known as s, p, d, f.

The atoms and ions inside the photospheres of stars create a great complexity in the rationalization of all possible electronic transitions. This complexity is based on the fine structure and hyperfine structure generated inside the multi-electron energy levels. Therefore the graphical representation is an excellent guidance for astronomers to quickly reveal the origin of the strongest lines in the recorded spectrum. Complete Grotrian diagrams representing all possible energy levels are not always needed, therefore certain selections of possible transitions are gathered in the so called partial Grotrian diagram, which fits well the needs for astrophysical interest. A generalized scheme of a Grotrian diagram is shown in Figure 3.2.

For allowed transitions the following selection rules are important:

- l quantum numbers (s, p, d, f) always change by +1 or –1. This way transitions occur here only between adjacent columns.
- The spin quantum number S ($\pm 1/2$) remains unchanged.

3.2.4 Spectroscopic Notation: The Term Symbol

For a correct interpretation of Grotrian diagrams and to analyze more profoundly the different possible transitions

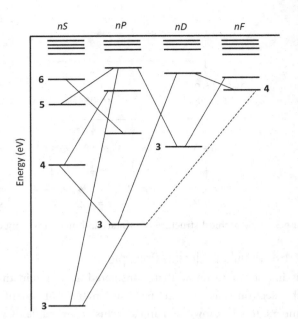

Figure 3.2 Generalized scheme of a Grotrian diagram

between the energy levels the term symbol is introduced. It is an abbreviated notation for a specific electronic state. Going from a system with one electron to multiple electron systems it is the total angular momentum that contributes to the energy of the system. The possible values of the total angular momentum create the different energy levels. The term symbol has a specific (but somewhat confusing) notation and is presented by Equation {3.3}:

$$^{(2S+1)}L_J \qquad \{3.3\}$$

where 2S+1 is defined as the *total spin multiplicity*, which gives the number of possible orientations of the spin angular momentum given by the total spin quantum number S. And 2S+1 is the mathematical multiplicity for $S = \sum M_S$. In practice this means that the term 2S, being twice the sum of the electron spins, is equivalent to the number of active valence electrons. Active means valence electrons that are in a position to become chemically active. Consequently, adding one to this number gives the multiplicity of the term. For diatomic or even polyatomic molecules the method becomes complicated and other methods can be used such as the MO (molecular orbital) method extended with the LCAO approach (linear combination of atomic orbitals) [16] [17]. The multiplicity 2S+1 is finally placed as a superscript on the left side of the ground term symbol L. Some examples of calculated multiplicity are presented in Table 3.2.

The ground term L stands for the *total orbital angular momentum* and is the coupling of all individual electronic angular moments. Together with the orbital angular

Table 3.2 Multiplicity calculation for some atoms

Atom	Electron configuration	Multiplicity
H	$1s$	$1+1 = 2$
He	$1s^2$	$0+1 = 1$
N	$1s^2 2s^2 2p^3$	$3+1 = 4$
O	$1s^2 2s^2 2p^4$	$2+1 = 3$
Ne	$1s^2 2s^2 2p^6$	$0+1 = 1$
Fe	$1s^2 2s^2 2p^6 3s^2 3p^6 4s^2 3d^6$	$4+1 = 5$
Ni	$1s^2 2s^2 2p^6 3s^2 3p^6 4s^2 3d^8$	$2+1 = 3$
Zn	$1s^2 2s^2 2p^6 3s^2 3p^6 4s^2 3d^{10}$	$0+1 = 1$

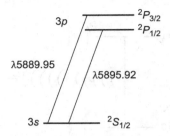

Figure 3.3 Doublet transitions of sodium D 1, 2 lines

momentum L the total spin angular momentum S, which couples all individual spin angular moments is involved in a final coupling, defined as the *total angular momentum* $J = L+S$, which is also called the Russell–Saunders coupling. First published in 1925 by the Canadian physicist Frederick Albert Saunders (1875–1963) and the American astronomer Henry Norris Russell (1877–1957), it is also known as *L–S* coupling. The symbol J is placed as a subscript (sometimes as a superscript) on the right side of the term symbol L. The values of J vary between the largest absolute value $L + S$ and the smallest being $L – S$.

As an example we calculate the term symbol for the energy levels of the well known sodium doublet emission lines of $\lambda 5889.95$ (D_1) and $\lambda 5895.92$ (D_2). From the table of Mendeleev we read that sodium has atomic number 11, so there are 11 electrons to be distributed over the orbitals. The electron configuration for neutral sodium (Na I) is $1s^2 2s^2 2p^6 3s^1$. According to Section 3.2.1 the energy levels $1s$, $2s$ and $2p$ are completely filled up with the maximum number of electrons. This is called a closed shell status, which means that those inner shell electrons do not contribute to the angular momentum. The single electron, which resides in the $3s$ orbital, has an individual angular momentum of $l = 0$, because it is an s orbital, hence $L = 0$. The electron has a spin momentum $s = 1/2$, so $S = 1/2$. For this state, called the ground state the total angular momentum $J = L + S = 0 + 1/2$ or $0 – 1/2$ gives $1/2$. When the single $3s$ electron goes to the next energy level, it arrives in a p orbital, hence the configuration $3p^1$. The superscript 1 means here is one electron in play. There the angular momentum is $l = 1$, because it is a p orbital, hence $L = 1$. The spin momentum of the electron $s = 1/2$, so $S = 1/2$. The possible values of J here are $1/2$ and $3/2$, which results in two energy levels.

According to Section 3.2.1 the values of L, by convention, do have a corresponding letter, so $L = 0$ stands for term symbol S. Important note: This has nothing to do with spin quantum momentum S! The term symbol P is given by $L = 1$, and $L = 2$ gives D, and so on. In the case of sodium the term symbols become: $^2S_{1/2}$ (ground state), $^2P_{1/2}$ (excited state 1) and $^2P_{3/2}$ (excited state 2).

In the ground state ($3s^1$) there is 1 electron so $S = 1/2$, in the excited state ($3p^1$) there is also one electron, which came from the $3s$ energy level, so S is also $1/2$, hence both states do have the same multiplicity $2S + 1 = 2$, which is a doublet! Transitions from the two excited states to the ground state give rise to the typical yellow doublet emission lines observed in the spectrum as demonstrated in Figure 3.3.

Parity of the term symbol can also be calculated. Total parity will only be odd with an odd number of electrons located in odd orbitals. So for sodium the ground state is even, excited state, which is here a p orbital is odd with an odd number of electrons i.e. 1. So both term symbols have odd parity. It can be indicated by the small o as a superscript to the right side of the term symbol as follows:

$^2P^o_{1/2}$ and $^2P^o_{3/2}$

Other indicators are subscripts to the right of the term symbol with the letter g or u.

But when we force an electron in neutral sodium by laser excitation from the $2p$ shell to go to the $3p$ or $4p$ level we get the configuration $1s^2 2s^2 2p^5 3s^1 3p^1$ or $1s^1 2s^2 2p^5 3s^1 4p^1$, then the term symbol parity becomes even.

3.2.5 The Hyperfine Structure

The splitting of energy levels in atoms, ions or molecules as a result of the interaction between the spin of the electron and the nucleus is known as the hyperfine structure. Consequently extra energy levels are generated with parallel spin state in a higher energy level and the anti-parallel spin state at the lower level. Such spin-flip transitions between the energy states can

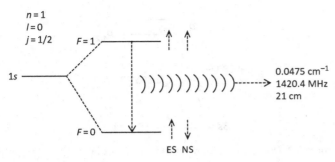

Figure 3.4 Hyperfine structure of H I

lead to absorption or emission and are generally three magnitudes weaker than those generated by the fine structure. The energy difference between the parallel and the antiparallel state is only 5.9×10^{-6} eV! Therefore the lifetime in the parallel state can easily reach 10^7 years! However, the abundance of hydrogen is extremely high, for instance statistically enough to enable significant radiation. In Figure 3.4 the hyperfine structure energy levels for H I are shown.

The "1s" ground state level of neutral hydrogen splits into two levels as a result of the changing arrangement of the electron spin, indicated as ES and the spin of the nucleus, abbreviated as NS. In quantum mechanical terminology, the sum of the angular orbital momentum L of the electron and the intrinsic spin angular momentum S of the electron is defined as the electron's total angular momentum J, which couples with the spin angular momentum of the nucleus I. The values for the total, the final angular momentum F, being the vector sum of J and I, are 0 and 1, resulting in the two energy levels shown in Figure 3.4. Due to the energy difference of 5.9×10^{-6} eV and according to Equation {1.5} the emitted radiation has a relatively long wavelength of ~21 cm. The corresponding frequency is 1420.406 MHz, lying within the UHF range which is also used by our mobile phones. This H I, also called "H 1" line, is an important subject for research not only by professional radio telescopes but also for amateur radio astronomers, for example to analyze the structure of galaxies. The 21 cm radio map of the Milky Way is the best known example of hyperfine structure recording.

3.3 The Schrödinger Equation

3.3.1 Preliminary Remarks

It was in the 1920s that the Austrian physicist and Nobel Prize winner Erwin Schrödinger (1887–1961) first formulated his "wave equation." His publication in the January 1926 edition of *Annalen der Physik* caused a revolution in the

scientific world of physics and chemistry. For spectroscopy this "wave equation" is very important because it describes how the quantum state of a physical system changes with time and further predicts the possible energy levels, introduced in Section 2.1 with the simple model of hydrogen. Its solution is called the "wave function Ψ" (Greek: psi) which, typical for quantum mechanics, gives just the *probability* of finding a particle at a certain location.

Unfortunately, this theory and the underlying models are highly abstract. On a first glance many of the involved equations may look surprisingly simple (e.g. Equation {3.6}). However, examined more closely some of the odd looking "variables" turn out to be "operators" substituting complex differential equations. Therefore a deeper understanding of this matter would require advanced knowledge of quantum mechanics and mathematics – this on a level which cannot be expected for most amateurs, the main target group of this book. Anyway the good message here is: For a meaningful, practical employment with astronomical spectroscopy this section can be skipped without losing the thread of the story. For those which are interested we introduce first the "wave function" with the quite easily understandable de Broglie model. Advanced students can follow some further explanations to the Schrödinger equation in order to gain a deeper understanding of the different models presented in Section 3.2 for multi-electron atoms.

3.3.2 De Broglie's Electron Wave Model

With de Broglie's model we can show why exactly a wave function is suited as a key to understand the different models explained in Section 3.2. The Danish physicist Niels Bohr (1885–1962) recognized that his famous atomic model was not able to explain all aspects. So a negative charged electron orbiting a positive atomic nucleus should radiate energy, that, however, does not happen. Moreover this system would consume energy this way and due to electrostatic attraction, the electron would inevitably spiral in to the core within $\sim 10^{-11}$ seconds.

The French physicist and Nobel Prize winner Prince Louis de Broglie (1892–1987) was a member of an illustrious aristocratic family. In 1923 as a student he had suggested that for a possible way out of this dilemma electrons may be considered as standing waves. He was probably inspired by the wave–particle duality of photons, which was already accepted at that time. It refers to the photoelectric effect discovered in 1887 by the German physicist Heinrich Hertz (1857–1894). In 1905 Albert Einstein (1879–1955) explained this effect by postulating the presence of "energy quanta," later defined as

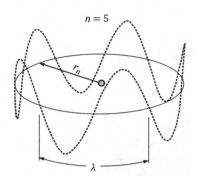

$n = 5$

Figure 3.5 De Broglie standing wave for main quantum number $n = 5$

photons. De Broglie's approach seemed to be plausible because standing waves require no energy, and further fit to the spectroscopic observations, exhibiting quantized wavelengths, which are related to accordingly quantized energies.

Figure 3.5 shows as an example the circular orbit of the quantized energy level with the main quantum number $n = 5$ and the corresponding radius r_n, according to the Bohr model.

It forms the zero line for the equal number of five stable, standing de Broglie waves, representing the electron. Their length must exactly fit the circumference, which must be an integer multiple of the wavelength. This way a clear relationship becomes obvious between the known orbital radius r_n and wavelength λ. This is generally expressed as:

$$2r_n\pi = n\lambda \qquad \{3.4\}$$

Strongly simplified this sinusoidal wave can now be described as a periodic wave function Ψ depending on the considered phase ϕ and with maximum amplitude A:

$$\Psi(\phi) = A\sin\phi \qquad \{3.5\}$$

3.3.3 The Schrödinger Equation and Astronomical Spectroscopy

In 1926, this model became finally the base for Erwin Schrödinger to develop a mathematical model, describing the status of an electron by the wave function Ψ. The mathematical derivations are deliberately restricted to a minimum as they are beyond the scope of this book. Only those equations and formulas are being discussed here which represent an important and relevant contribution to a better understanding of astronomical spectroscopy. In this section we will make a small but necessary digression about one of the most important scientific equations of the twentieth century in the field of astrophysics: the Schrödinger equation (SE).

With the exception of Equations $\{3.12\}$ and $\{3.18\}$ this short section assumes the reader has a more advanced knowledge of quantum mechanical terminology, models and formulas.

Before starting, a small digression on the use of operators in quantum mechanics is presented. Properties of a physical system are characterized by its observables, for instance energy, position, momentum and angular properties. Quantum states of a system are defined by their mathematical representation, the wave function Ψ. For every observable a linear *operator* can act on the wave function. They are called Hermitian operators, named after the French mathematician Charles Hermite (1822–1901) and change one function into another. An example is the operator for the total energy, called the Hamiltonian $\hat{H}$. It was named after the Irish physicist and astronomer William Hamilton (1805–1865). Another frequently used operator is the Laplacian ∇^2 or Δ, named after the French mathematician Pierre-Simon Laplace (1741–1827). It is a differential operator which develops Cartesian or polar coordinates and is used by the kinetic energy operator T and by the angular momentum operator L. These types of operators *describe* the observable and can be considered as type I operators. A second type of operator causes a *transformation* in position and/or in time. An example of this type is the parity operator P which induces an inversion. The measurable values of an operator are called the eigenvalues and are strictly determined by the eigenstate of the operator.

3.3.4 Time-independent and Time-dependent Forms

The most general form of a *time-independent form* (TISE) is given by Equation $\{3.6\}$:

$$\hat{H}\,\Psi = E\Psi \qquad \{3.6\}$$

where $\hat{H}$ represents the Hamiltonian operator which is the physical variable of the energy of the system i.e. the sum of its kinetic and potential energy. In this form the Hamiltonian is not time-dependent. The wave function Ψ, represents here the stationary state, and E is the energy of the state, the eigenvalue, to be interpreted as a characteristic property of the system corresponding to the wave function Ψ. The practical interpretation of the time-independent form considers the probability distributions not to change with time. The stationary state equals that of a standing wave.

The Schrödinger equation exists also in another form. Besides the time-independent form there is also a *time-dependent form* (TDSE). The general form of the TDSE for

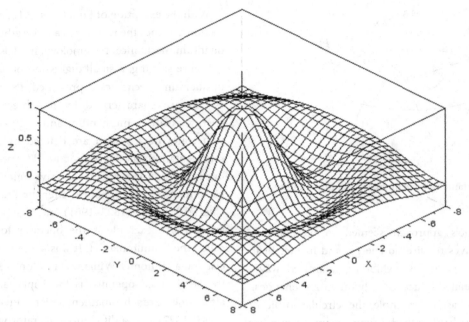

Figure 3.6 Course of a sinc function in 3D

a particle without external interaction, the so called "time evolution" form is presented by Equation {3.7}:

$$ i\hbar \frac{\partial \Psi}{\partial t} = \hat{H}\Psi \qquad \{3.7\} $$

Where i is the imaginary number, $\hbar$ (pronounced as h bar) is the reduced Planck's constant $\hbar = h/2\pi$ (also called Dirac's constant), and t is time.

For a physical system without interaction between radiation and matter, for instance without interference of electric and magnetic fields, it is possible to have a solution for the Schrödinger equation. Here we can consider the system being in consecutive time-independent stationary phases.

The presence of an electric or magnetic field causes a perturbation. As a result transitions can occur between energy levels. The Hamiltonian operator then becomes time-dependent. In perturbation theory one defines an unperturbed Hamiltonian as $\hat{H}^0$ and a perturbed one as $\hat{H}'$. As an approximation we assume the difference between $\hat{H}^0$ and $\hat{H}'$ to be small. The total $\hat{H}$ is given by Equation {3.8}:

$$ \hat{H} = \hat{H}^0 + \hat{H}'(t) \qquad \{3.8\} $$

The solution for the time-dependent Schrödinger equation of any state can be generalized as:

$$ \Psi_n(x, t) = \Psi_n(x)e^{-iE_n t/\hbar} \qquad \{3.9\} $$

To calculate the probability for transition from energy level m to n we define the energy difference between the two levels expressed as ω_{mn} (with $\omega = 2\pi v$ as the angular frequency).

Substituting Equation {3.8} into {3.9} and combined with the wave function expansion based on Equation {3.10} for two energy levels m and n gives the following term for the *proportionality* $P_{m \to n}(t)$ concerning the absorption $m \to n$:

$$ P_{m \to n}(t) \approx \frac{\sin^2\left\{\frac{1}{2}(\omega_{mn} - \omega)\right\}t}{(\omega_{mn} - \omega)^2} \qquad \{3.10\} $$

The result of Equation {3.10} looks like a cardinal sine or sinc function. Figure 3.6 displays the course of such a sinc function in 3D.

The sinc term in Equation {3.10} reaches its maximum as the variable $\omega_{mn} - \omega$ is approaching zero, but falls quickly as the variable reaches values $>$ or $<$ zero. Therefore the probability of an absorption is highest when $\omega_{mn} = \omega$ or otherwise $hv_{mn} = hv$. For all values moving away from zero, the probability is lowest and no absorption occurs. It can be seen as an all or none phenomenon.

3.3.5 Hydrogen

The Schrödinger equation for a particle with mass m is given in Equation {3.11}:

$$ -\frac{\hbar^2}{2m}\nabla^2\Psi + V\Psi = i\hbar\frac{\delta\Psi}{\delta T} \qquad \{3.11\} $$

The term containing the Laplace operator here in the form of ∇^2 or nabla^2 represents the kinetic energy. The second term $V\Psi$ represents the potential energy. The hydrogen atom consists of one proton in the nucleus and one electron. The potential energy is given by Equation {3.12}:

$$V(r) = \frac{-Ze^2}{4\pi\varepsilon_0 r} \qquad \{3.12\}$$

where Z is the charge of the nucleus, which is 1 for hydrogen and e represents the charge of the electron. The permittivity of the vacuum is indicated by ε_0 and r is the electron–nucleus distance.

Substituting Equation {3.12} in Equation {3.11} gives the Hamiltonian operator Equation {3.13}:

$$\hat{H} = -\frac{\hbar^2}{2\mu}\nabla^2 - \frac{Ze^2}{4\pi\varepsilon_0 r} \qquad \{3.13\}$$

The μ introduced here represents the reduced mass, defined as $\frac{mM}{m+M}$ where m is the mass of the electron and M refers to the mass of nucleus.

This hydrogen Schrödinger equation is not easy to solve. Therefore the *angular* and *radial* parts have to be separated by introducing a three-dimensional model of the SE. Solving both parts then gives the allowed values of angular momentum and energy.

Substituting spatial x-, y-, z-coordinates of the Laplacian operator ∇^2 in Equation {3.11} gives Equation {3.14}, which represents the transformation of the SE according to the spherical coordinates system:

$$-\frac{\hbar^2}{2\mu r^2}\frac{\partial}{\partial r}\left(r^2\frac{\partial\Psi}{\partial r}\right) - \frac{\hbar^2}{2\mu r^2}\frac{1}{\sin\theta}\frac{\partial}{\partial\theta}\left(\sin\theta\frac{\partial\Psi}{\partial\theta}\right)$$
$$-\frac{\hbar^2}{2\mu r^2}\frac{1}{\sin^2\theta}\frac{\partial^2\Psi}{\partial\phi^2} + V(r)\Psi = E\Psi \qquad \{3.14\}$$

After recombination and introducing the radial wave function $R(r)$ together with the angular wave function $Y(\theta,\phi)$ in a product wave function form presented as Equation {3.15} the separation of variables can be performed:

$$\psi(r,\theta,\phi) = R(r)Y(\theta,\phi) \qquad \{3.15\}$$

The result of the separation yields the radial equation {3.16} and {3.17} the angular equation with l defined as the angular momentum quantum number, also known as the orbital quantum number or azimuthal quantum number:

$$\frac{1}{R}\frac{\partial}{\partial r}\left(r^2\frac{\partial R}{\partial r}\right) - \frac{2\mu r^2}{\hbar^2}V(r) + \frac{2\mu r^2}{\hbar^2}E = l(l+1) \qquad \{3.16\}$$

$$\frac{1}{\sin\theta}\frac{\partial}{\partial\theta}\left(\sin\theta\frac{\partial Y}{\partial\theta}\right) + \frac{1}{\sin^2\theta}\frac{\partial^2 Y}{\partial\phi^2} = -l(l+1) \qquad \{3.17\}$$

The solution of the radial equation {3.16} gives Equation {3.18} which represents the energy levels of the hydrogen atom and introduces n as the principal quantum number:

s	$l=0$	$m_l=0$	
p	$l=1$	$m_l=0, \pm1$	
d	$l=2$	$m_l=\pm1, \pm2$	
d_{z2}	$l=2$	$m_l=0$	

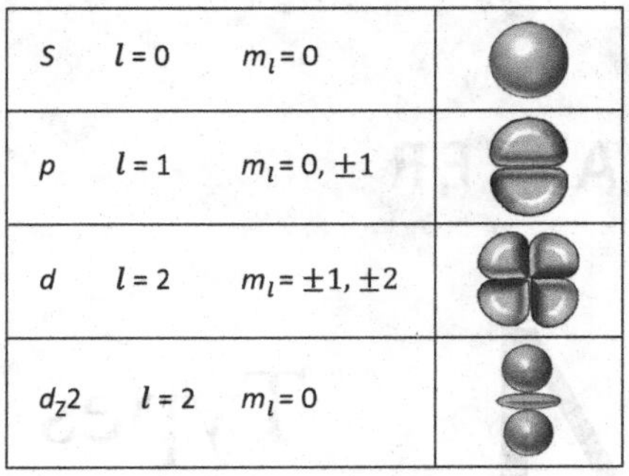

Figure 3.7 Shapes of s, p and d orbitals

$$E_n = -\frac{\mu Z^2 e^4}{8\,\varepsilon_0^2\,\hbar^2 n^2} = -\frac{13.6\,\text{eV}}{n^2} \qquad \{3.18\}$$

The ground level ($n = 1$) of the hydrogen atom gives, according to Equation {3.18}, an ionization energy of 13.6 eV, which correlates with the convergence limit of the Lyman series at 910 Å (see Section 2.4.4).

The solution (Equation {3.20}) of the angular equation {3.17} goes over the separation of the azimuthal (ϕ) and the polar (θ) part by introducing the product defined according to Equation {3.19}:

$$Y(\theta,\phi) = \Theta(\theta)\Phi(\phi) \qquad \{3.19\}$$

The final solution to the angular equation is presented by Equation {3.20}:

$$Y_{l,m_l}(\theta,\phi) = A\cdot P_{m_l}{}^l(\cos\theta)\cdot e^{im_l\phi} \qquad \{3.20\}$$

Here, m_l represents a quantum number giving the possible values of the wave function and is called the magnetic quantum number. With A as a normalization constant given by Equation {3.21}:

$$A = \left[\frac{(2l+1)(l-m_l)!}{4\pi(l+m_l)!}\right]^{1/2} \qquad \{3.21\}$$

The result is a Legendre polynomial describing the different shapes of the wave function, which are also called the spherical harmonics. The shapes of s, p and d orbitals are presented in Figure 3.7.

To summarize we conclude that besides the prediction of the quantized atomic/molecular energy levels and the probability to find an electron at a certain location, the Schrödinger equation introduces n as the principal quantum number, l as the angular momentum quantum number, m_l as the magnetic quantum number and the atomic/molecular orbitals based on a three-dimensional wave function model.

4 Types and Function of Dispersive Elements

4.1 Physical Principle of Dispersion

4.1.1 Preliminary Remarks

The element of dispersion can be considered as the working heart of a spectrograph. Dispersion of light is *a conditio sine qua non* for analyzing the incoming light and is defined as the separation of light by refraction or diffraction. The understanding of either of those two dispersive mechanisms is indispensable for anyone who wants to construct a spectrograph or to make a choice in the type of spectrograph to be used in specific projects. Therefore it is considered to be essential reading for a more profound insight in this phenomenon.

4.1.2 Dispersion by Refraction

One of the best examples to illustrate the physical principle of this type of dispersion is the appearance of a rainbow. Figure 4.1 shows a photograph which Marc Trypsteen took out of his car on a rainy day, in it one can see the primary and the secondary arcs. Between both arcs is a dark band, the so called Alexander band, named after the Greek philosopher Alexander of Aphrodisias, who first reported the phenomenon in the year 200 CE. Much later, in 1637 the mathematical science behind the rainbow was published and clarified by the French mathematician Rene Descartes (1596–1650). It was partly based on the law of refraction, known as Snell's law, which was named after the Dutch mathematician and astronomer Willebrord Snel van Royen, better known by his Latin family name Snellius (1580–1626). With the evolution

in physics, new mathematical methods were proposed by Christiaan Huygens (1629–1695) in 1670, Augustin Jean Fresnel (1788–1827) in 1815, Sir George Biddell Airy (1801–1892) in 1838 with the Airy theory, Ludvig Lorenz (1829–1891) in 1890, Gustav Mie (1869–1957) in 1908 with the Lorenz–Mie theory. Nowadays, computer simulation programs such as *Airysim*, *Bowsim* [84] and *Mieplot* [18] are used to study rainbows. A rainbow can be seen as a naturally generated continuous spectrum. The typical observed colors of a rainbow are red, orange, yellow, green, blue, indigo and violet.

This shifting of colors is called dispersion, and in the case of a rainbow the causal mechanism is refraction. The additional internal reflection inside of the water droplets generates the primary arc. A second internal reflection creates a secondary arc and as a result the ordering of the colors is

Figure 4.1 Dispersion by refraction: rainbows
(Credit: Marc Trypsteen)

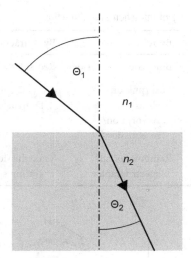

Figure 4.2 Snell's law of refraction

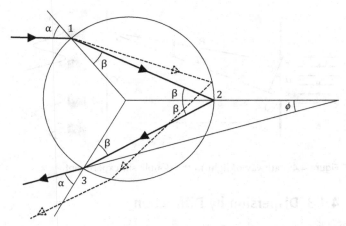

Figure 4.3 Pathway of a beam of light in an (idealized) water droplet, generating a rainbow

reversed. The zone between the two arcs contains rain droplets which do not reflect any light in the direction of the observer, hence the observed dark band.

The border of the water droplet forms an interface between two media, air and water, which each have their own index of refraction. The light rays are bended and refracted when they pass from the medium air to the medium water and vice versa.

Knowing the refractive indices from air to water of red and blue light respectively we can calculate the angle of refraction for each color. Therefore we use Snell's law (Figure 4.2), also known as the law of refraction and defined by the general equation {4.1}.

$$n_1 \sin \theta_1 = n_2 \sin \theta_2 \qquad \{4.1\}$$

n_1 and n_2 are the refractive indices of the different media and θ_1 and θ_2 the angles of the incident and refracted beam respectively with respect to the surface normal.

The pathway of the light through an idealized raindrop is illustrated by Figure 4.3. In the case of a rainbow the visible light is refracted at point 1 upon entering the water droplet and after internal reflection at point 2 inside the droplet, it leaves the droplet as a dispersed light bundle at point 3.

Applying Equation {4.1} to the situation of an incident beam of light on the water droplets and relative to the refractive index of the surrounding medium which is ~1 for air, gives the equation of Snell's law in another form resulting in Equation {4.2}:

$$\sin \alpha = n \sin \beta \qquad \{4.2\}$$

with α as the angle of incidence on the water droplet and β is the angle of refraction inside the water droplet as indicated in

Figure 4.3. Also from Figure 4.3 we can calculate the angle ϕ which gives us the maximum value for the angle of the rays of light leaving the droplet:

$$\phi = 4\beta - 2\alpha \qquad \{4.3\}$$

Given the air to water refractive indices for red and blue light as 1.33 and 1.34, we calculate for $\phi_{red} = 42.4°$ and for $\phi_{blue} = 40.6°$ respectively. This means that, depending on the angle of observation, we can only see red light at 42° and blue light at 40°. The primary color band of a rainbow is therefore only visible between 40° and 42°!

To summarize, the three conditions necessary to see a rainbow:

- The observer must always have the *light source* (e.g. Sun, Moon) *at their back.*
- The height of the light source (Sun, Moon . . .) *must not be higher than 42°.*
- In front of the observer *water drops are in free fall or soaring* (rain, mist, fog, location near waterfalls).

The mechanism of dispersion by refraction finds its origin in the interaction of light with matter, in this case water droplets. The different beams of light interact with the electrons of the water molecules. The electron configuration of two hydrogen atoms ($1s^1$) with one oxygen atom ($1s^2 2s^2 2p^4$) to form one water molecule mounts up to a total of 10 electrons. Blue light with a wavelength of 4500 Å causes a push/pull on those 10 electrons resulting in ~200 trillion vibrations per second (Hz) more than red light does with its wavelength of 6670 Å. After this absorption and re-emission process the speed through the water droplets of the blue light is slower than that of red light. As a result blue light is more refracted than red light, hence the change of their individual direction.

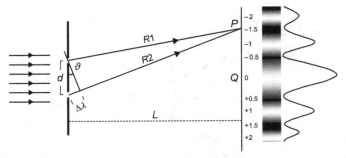

Figure 4.4 Pathway of light in the double slit experiment

4.1.3 Dispersion by Diffraction

Another way to shift wavelengths is to get dispersion by diffraction. It was discovered by the English physician Thomas Young (1773–1829) who carried out the famous double slit experiment. Therein a monochromatic light beam goes through a screen with two slits. The expected result was to see two lines on the viewing screen. But instead multiple lines were observed. In fact the light spreads out and the only possibility to understand what happened is to consider light as a wave instead of a particle. This phenomenon is called diffraction and the dark and bright fringes with specific spacing, is called an interference pattern. Historically it was the first experiment to show the wave nature of light.

In Figure 4.4 light goes through a double slit and is projected on a screen. The difference in the pathway between $R1$ and $R2$ is indicated as $\Delta\lambda$. The projection point on the screen is P.

When $R1 - R2 = \Delta\lambda$ then $\Delta\lambda = d \sin\theta$. If we consider now the distance L to the screen to be very large compared to the distance d between the two slits, we can assume that light rays $R1$ and $R2$ are approximately parallel to each other or in other words the slit equals a point source of light. The interference pattern projected on the screen is schematically presented on the right side of Figure 4.4. The distance PQ gives for each band the projection point P with respect to the center of the slit Q. The generated pattern appears as bright spots (antinodes or "peaks") when $\Delta\lambda$ equals whole numbers of λ, ($m = 1, 2, 3, \ldots$), which is called constructive interference. When the interference is destructive ($m = 0, 0.5, 1.5, \ldots$) in Figure 4.4 dark spots ("troughs") or nodes appear on the screen. In the case of constructive interference we can define $\Delta\lambda = m\lambda$ and for destructive interference $\Delta\lambda = (m + 0.5)\lambda$.

As a result the according equations appear as:

Antinodes : $m\lambda = d \sin\theta$ {4.4}

Nodes : $\left(m + \dfrac{1}{2}\right)\lambda = d \sin\theta$ {4.5}

Table 4.1 Summary of refraction and diffraction

Dispersion	By refraction	By diffraction
Bending	Blue side > Red side	Red side > Blue side
Causing mechanism	Push/pull on electrons Absorption/ re-emission	λ versus slit opening Dimension difference
Examples	Rainbow, prism spectrographs	Transmission/reflective grating spectrographs

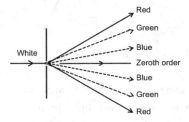

Figure 4.5 Diffraction of white light

The point of projection relative to the center of the slit for the interference fringes can be calculated from Equations {4.6} and {4.7} assuming the distance PQ being much smaller than L:

Constructive interference : $PQ = \dfrac{mL\lambda}{d}$ {4.6}

Destructive interference : $PQ = \dfrac{(m + \frac{1}{2})L\lambda}{d}$ {4.7}

According to Equations {4.4} and {4.5} the sine of θ is proportional to λ. As a consequence and also demonstrated by Figure 4.5, red light is diffracted more than blue light, which is just the opposite in the case of dispersion generated by refraction!

The mechanism of interaction for the diffraction phenomenon is based on the difference of the dimensions between the opening of the slit and the wavelength of each ray. As the wavelength on the blue side of the spectrum is shorter, passing through the slit is a priori easier than in the case for wavelengths of the red side. Hence the blue side of the generated spectrum is less deviated.

Table 4.1 shows a summary of dispersion by refraction and diffraction.

4.2 The Dispersive Principle of Prism Spectrographs

This type of spectrograph uses a prism as dispersive element. The spectrum is generated by refraction, which mechanism is comparable to the generation of the rainbow and is projected within the so called angle of dispersion. The prism used is a dispersive equilateral prism made of a glass type with a considerable dispersion power. Internal angles are 60°.

In contrast to these *dispersive* prisms there are also the *redirecting* prisms. They have the typical 45° – 90° – 45° geometry of the isosceles triangles. The most important factor that determines the dispersive power of the prism is the quality of the glass.

4.2.1 Specific Glass Types for Prism Spectrographs

Table 4.2 gives an overview of a broad collection of commercially available glass types. The best way to become familiar with the different types is to check the glass number or international glass code (GC). In Table 4.2 it is positioned in the second column and presented as a six digit number. The first three numbers, indicated as GCA, give the specification on the refractive index of the glass at a reference wavelength, for instance Fraunhofer D or d line, situated in the yellow part of the visible spectrum. The last three numbers, indicated as GCB, correspond to the Abbe number V_d. In some catalogs of glass manufacturers three extra numbers are added which correspond to the density. The interpretation for a given glass code or number is as follows:

Glass code or number: $\underbrace{517}_{GCA}\underbrace{642}_{GCB}$

where GCA and GCB are glass code A and B respectively. With Equation {4.8} the corresponding values of the refractive index n_d and Abbe number V_d can be calculated:

$$n_d = \frac{GCA}{1000} + 1, V_d = \frac{GCB}{10} \qquad \{4.8\}$$

The Abbe number is an indicator of optical dispersion corresponding to the glass type and is simultaneously a measure of the degree of chromatic aberration. It is named after the German physicist Ernst Abbe (1840–1905) and is calculated according to the following form and most used equation {4.9}:

$$V_d = \frac{n_d - 1}{n_F - n_C} \qquad \{4.9\}$$

with n_d, n_F, n_C and n_h the refractive indices of the glass at Fraunhofer d, F and C lines and in addition to the h line. Important note here on the convention: The spectral line catalogues of glass manufacturers indicates the violet line h

not as the Fraunhofer h line, but as the violet mercury line. d corresponds to λ5876 (He I), F to λ4861 (Hβ), C to λ6563 (Hα) and h to λ4046.6 (violet Hg line). Originally the Abbe number, V_D was calculated based on the Fraunhofer D line, which is a sodium absorption at λ5893 (blend of D 1 and D 2). Later the reference wavelength was changed from the Fraunhofer D to the Fraunhofer d (or D 3) line, which represents the He I line at λ5876. Glass manufacturers generally use the latter as the reference wavelength.

The value $n_F - n_C$ is the principal or main dispersion. Dividing the main dispersion by the term $n_d - 1$ is defined as the relative dispersion. As a result of Equation {4.9} the Abbe number is proportional to the reciprocal value of the relative dispersion. The higher the dispersion the lower the Abbe number and vice versa.

It is clear that choosing a glass type for a prism to be used in a spectrograph, the Abbe number has to be low and for glasses used for reading or to manufacture a good quality apochromatic refractor telescope, the Abbe number must be high. Hence the use of fluorite or ED glass for quality telescopes, which can have Abbe numbers even higher than 90. The different types of glass are roughly divided in two groups. The first group is called *crown glass* and has an Abbe number higher than 55. The second group is called *flint glass* with an Abbe number lower than 55.

To calculate the refractive indices of all available types of glass types at other wavelengths a general dispersion formula is used. In history many efforts have been made to find the best fitting formula. Generally accepted in the optical industry as the standard dispersion formula is the three-term Sellmeier formula, presented by Equation {4.10}. It was formulated by the German physicist Wolfgang Sellmeier in 1871 as a more accurate form of the Cauchy equation:

$$n^2 - 1 = \frac{A_1\lambda^2}{\lambda^2 - B_1} + \frac{A_2\lambda^2}{\lambda^2 - B_2} + \frac{A_3\lambda^2}{\lambda^2 - B_3} \qquad \{4.10\}$$

where A_1, A_2, A_3, B_1, B_2 and B_3 are the so called Sellmeier constants which correspond to the glass type and are indicated by the glass manufacturer, n is the refractive index and λ is the wavelength expressed in [µm]. Accuracy on the calculated refractive index is $\pm 5 \times 10^{-6}$.

4.2.2 Minimum Angle of Deviation and Angular Dispersion

Prism spectrographs exist in two different concepts: The *objective prism spectroscope* and the *classical prism spectrograph*. Both arrangements use a prism installed at minimum angle of deviation. The spectrographic arrangement to achieve the position of the prism at minimum angle of deviation is typified by the symmetry between entering and exiting angles i and r.

Inside the prism, the pathway of the light runs parallel to the base of the prism as shown in Figure 4.6.

Here α is the prism angle or apex, i the angle of incident light, s and t are internal angles of refraction and δ is the angle of deviation. Angles u and v are used for geometric calculations. The relationship of the minimum angle of deviation and the refractive index is determined by Equation {4.11}, known as the prism equation:

$$n \sin\left(\frac{\alpha}{2}\right) = \sin\left(\frac{\alpha + \delta}{2}\right) \qquad \{4.11\}$$

By measuring the angle of entry corresponding to minimum deviation the index of refraction of the prism can be calculated with Equation {4.11}, The index of refraction of the prism determines the resulting separation of light rays in its

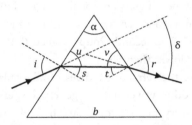

Figure 4.6 Pathway of light at minimum deviation angle through an equilateral prism

different wavelengths, which is called angular dispersion. Figure 4.7 shows the arrangement for the classical prism spectrograph. The light rays coming from the slit run parallel after passing through a collimating lens, then are refracted by the prism, generating a spectrum, pass afterwards through a focusing lens and are finally projected on the sensor of a camera device, where the spectrum can be observed.

The example of the calculation of the angular dispersion in the range violet color of the mercury line (h line at 4046.6 Å) to the red color of the Hα (Fraunhofer C line at 6563 Å) of an SF2 glass type equilateral prism with apex angle of 60° and an angle of incidence of 45° gives an idea of the projected dimension of the observed spectrum. Therefore we need to calculate the angles of refraction r of Hg and Hα by using Snell's law for the pathways of light indicated in Figures 4.6 and 4.7. The corresponding refractive indices are taken from Table 4.2. First we determine the angle of entry s for both wavelengths by application of Snell's law as in Equation {4.12}:

$$n_{\text{glass}} \sin s = n_{\text{air}} \sin i \rightarrow s = \sin^{-1}\left(\frac{n_{\text{air}} \sin i}{n_{\text{glass}}}\right) \qquad \{4.12\}$$

$$\text{for Hg} \rightarrow s_{\text{Hg}} = \sin^{-1}\left(\frac{1 \cdot \sin 45°}{1.68273}\right) = 24.85°$$

$$\text{for Hα} \rightarrow s_{\text{Hα}} = \sin^{-1}\left(\frac{1 \cdot \sin 45°}{1.64210}\right) = 25.51°$$

(Note, in Equations {4.12} and {4.13} $\sin^{-1}$ is just another kind of notation for arc sin!)

The next step is the determination of both exit angles r_{Hg} and $r_{\text{Hα}}$ again with Snell's law for the exit rays. From

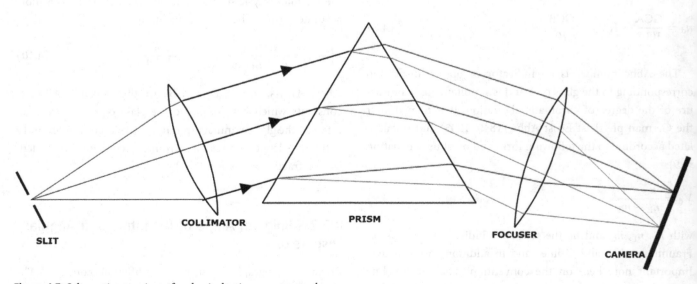

Figure 4.7 Schematic overview of a classical prism spectrograph arrangement

Table 4.2 Overview of glass types and corresponding numbers

Glass type	number	n_F	n_d	n_C	n_h	Common name
BaK4	569560	1.575912	1.56883	1.56575	1.58614	Barium crown
BK7	517642	1.522374	1.51680	1.51432	1.53024	Borosilicate crown
F2	620364	1.63208	1.62005	1.61506	1.65087	Flint
Fluorite	488704	1.49227	1.48750	1.48535	1.49894	Fluorite CaF_2 crown
FPL53	439950	1.44195	1.43875	1.43733	1.44645	ED
Fused silica	458677	1.46317	1.45849	1.45640	1.46966	Fused quartz
SF2	648338	1.66124	1.64769	1.64210	1.68273	Dense flint
SF6	805254	1.82783	1.80518	1.79608	1.86506	Dense flint
SF18	722293	1.73901	1.72151	1.71437	1.75640	Dense flint

Figure 4.6 we see that the angle $v = 30° + s$ and also that the angle $t = 90° - v$.

$$n_{air} \sin r = n_{glass} \sin t \rightarrow r = \sin^{-1}\left(\frac{n_{glass} \sin t}{n_{air}}\right) \quad \{4.13\}$$

$$\text{for Hg} \rightarrow r_{Hg} = \sin^{-1}\left(\frac{1.68273 \cdot \sin 35.15°}{1}\right) = 75.65°$$

$$\text{for H}\alpha \rightarrow r_{H\alpha} = \sin^{-1}\left(\frac{1.64210 \cdot \sin 34.49°}{1}\right) = 68.41°$$

Finally the angular dispersion of the light is $75.65° - 68.41° = 7.24°$ for the wavelength range from $\lambda4046.6$ to $\lambda6563$. By changing the angle of incidence and the glass type the optimal angular dispersion can be calculated.

An example of such type of prism spectrograph is the BESOS spectrograph, made by the CAOS spectroscopy group in 2010 [61]. The central dispersive element consists of an equilateral SF18 glass type 60° prism equipped with an MgF_2 reflective coating.

The second type of prism spectrograph is the objective prism spectroscope arrangement. Therein the prism is placed in front of a lens or telescope objective. The light rays of the point-shaped stars travel through the prism and after dispersion multiple spectra are registered at the same time. From a historical perspective this was the type of spectrograph used by the American astronomers Edward Pickering (1846–1919) and Annie Jump Cannon (1863–1941). The technique of objective prism spectroscopy facilitates large-scale surveys of bright stars and was carried out from 1886 to 1949. The work of Pickering and Cannon resulted in the famous complete Henri Draper catalogue and its extensions on the spectra of

Figure 4.8 Objective prism (21 cm diameter, 50 to 5 mm thickness slope, 12° tilt angle) mounted on top of an 11 cm f/5 Newtonian telescope.
(Credit: Mike Harlow (British Astronomical Association))

different star types, it was also the inspiration for the Harvard classification system of stellar spectra, still used today [1]. At that time the technique was called object prism spectrography as the spectra have been recorded on photographic plates. Sometimes up to four prisms were installed in front of the telescope objective.

In case the prism is smaller than the telescope front opening a shielding plate can be used to prevent direct stray light entering the telescope. A typical drawback of this setup in practice is the angle between the optical axis of the telescope and the aimed object.

4.2.3 Resolving Power: A Measure of Performance

When two monochromatic lines of comparable intensity are fully separated from each other and can be defined as λ and $\lambda+\Delta\lambda$ respectively, then the *resolving power R*, also called the resolvance, is defined by Equation {4.14}:

$$R = \frac{\lambda}{\Delta\lambda} \qquad \{4.14\}$$

where λ is the wavelength and $\Delta\lambda$ the wavelength interval defined as the *spectral resolution*. It is the smallest wavelength interval that can be distinguished and is expressed in wavelength units as Å or nanometers. In contrast to the spectral resolution $\Delta\lambda$, the resolving power R is a dimensionless quantity. For example, the resolving power needed for a spectrograph to resolve the sodium D1 and D2 lines at $\lambda5895.92$ and $\lambda5889.95$ respectively is given by applying Equation {4.14}:

$$R = \frac{5889.95}{5.97} = 987$$

In practice this means that we need a spectrograph reaching a value for R of ~1000 to resolve the sodium doublet. This method is commonly used as a standard criterion for the resolving power of a spectrograph.

In the case of a prism used as the main dispersive element in a spectrograph we can determine the R value for the prism by calculating the relation between the prism, Equation {4.11}, and the full width dispersion of the prism. This differentiating mathematical procedure results in the important Equation {4.15}, defining the capacity for the prism to separate two adjacent wavelengths:

$$R = \frac{\Delta n}{\Delta\lambda} b \qquad \{4.15\}$$

where n is the refractive index, λ the wavelength and b the base length of the prism (in mm), and $\Delta n/\Delta\lambda$ represents the dispersion. The theoretical calculated value of R is, according to Equation {4.15}, proportional to the dispersion and the base length of the prism. Hence the idea to install multiple prisms positioned in angle of minimum deviation to improve the resolving power. Of course other factors, such as the slit width, telescope parameters and pixel size of the camera, influence the theoretical calculated value of R. Nevertheless the value of R, calculated this way, gives a first directive of the achievable performance. For example, by applying Equation {4.15} we can determine an average value for R for the Fraunhofer lines from F to d of an SF2 glass type prism with a base length of 40 mm:

$$R = \frac{1.66124 - 1.64769}{5876 - 4861} \times 10^7 \times 40 = 5157$$

The calculated resolving power of 5157 corresponds to a resolution $\Delta\lambda$ of 1.14 Å in the yellow wavelength range based on the He I line at $\lambda5876$.

4.2.4 Practical Applications of Prisms Today

The use of prisms as the main dispersive element in spectrographs is very limited nowadays. The main reasons are the limited resolving power, the weight of the prism and the strongly nonlinear spectral output. The main application for using a prism in spectrographs is as a cross dispensing element in high resolution Echelle spectrographs and in combination with gratings in grisms (grating–prism). The advantage of their high transmission efficiency – up to ~98% – is the reason why they are sometimes used in low resolution professional spectrographs. They are very well suited for studying very faint objects.

4.3 The Dispersive Principle of Grating Spectrographs

A grating is an optical element used for the dispersion of light. It is composed of a glass plate with parallel lines or grooves on its surface and is covered with a reflecting aluminum or photosensitive coating. Gratings exist in two different types, transmission or reflection. With the former type incident and diffracted light rays are at opposite sites, with the latter they are at the same side as shown in Figure 4.9.

4.3.1 The Grating Equation

The physical principle of the dispersion by a grating is diffraction as discussed in Section 4.1. The spacing of the lines or grooves in the grating interacts with the different wavelengths

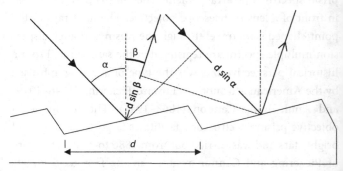

Figure 4.9 Reflection grating

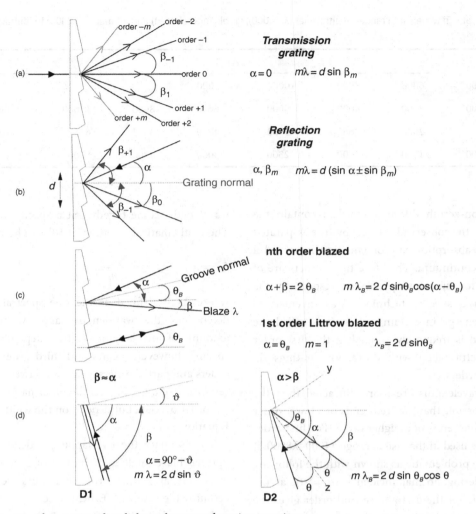

Figure 4.10 Grating types with corresponding light pathways and grating equations

of the incident light. As a result of this difference in dimensions an interference pattern is generated with alternate constructive and destructive character, ending up with the so called different diffraction orders, indicated in the general grating equation {4.16} and schematically shown in Figure 4.4.

$$m\lambda = d(\sin\alpha \pm \sin\beta) \qquad \{4.16\}$$

where m is the order of diffraction, which can have the integer values $-m$, 0, $+m$. The distance d between adjacent lines or grooves is also the reciprocal of the *grating constant*, which is expressed as the number of lines per mm (L mm^{-1}). The wavelength of the light is λ, α the angle of incident light and β the angle of diffraction typified by the order number. Both angles are measured relative to the grating normal or the grating perpendicular (dashed line) as shown in Figure 4.9. When both rays are on the same side of the normal for β the

(+) sign is used, when on opposite sides of the normal then the (−) sign is used conventionally.

Depending on the type of grating used, the orientation of the angle of incidence of the light, the shape of the grating grooves or facets it is possible to illustrate the physical meaning by calculating the corresponding mathematical result of Equation {4.16} for each situation. Most frequently used configurations in astronomical spectrographs are presented in Figure 4.10.

As seen from Figure 4.4 the energy distribution is spread out over the different orders. The zeroth order contains all wavelengths and is furthermore just as interesting as a reference point for a very rough wavelength calibration with slitless applied transmission gratings (Section 8.1.2). Furthermore, for slit spectrographs the sampling S can roughly be determined with a direct recording of the zeroth order slit image (Section 6.3.5).

Table 4.3 Overlapping ratio in wavelength ranges of ultraviolet ($\lambda < 4000$), visible (gray shaded) and infrared ($\lambda > 8000$) for diffraction orders 1 to 4 of a grating

Order m	λ (Å)								$m \rightarrow m+1$
1	4000	6000	8000	10000	12000				
2	2000	3000	4000	5000	6000	8000	10000	12000	1/2
3	1333	2000	2667	3333	4000	5333	6667	8000	2/3
4	1000	1500	2000	2500	3000	4000	5000	6000	3/4

However, the non-zeroth diffraction orders contain the information about the observed object by the separated wavelengths of the absorption and/or emission lines and a specific run of the continuum. Therefore these orders are of principal interest for spectroscopy. With respect to the energy distribution a specific technique was invented to concentrate more energy in certain orders. It is called the blaze technique and is frequently applied in the first order spectrum. But spectra taken with diffraction gratings do show overlapping orders.

As the longer wavelengths are more diffracted than the shorter ones, it is possible that the "red" ends of a lower order overlap with the "blue" ends of a higher order. For first order blazed gratings to be used in the visible range $\lambda 4000$–$\lambda 8000$, it is fortunately not so problematic as shown with the following calculation. Considering normal incidence ($\alpha = 0$), at the same deviation angle for the first and second order gives the following solution for the grating equation:

$$m_1 \, \lambda_{m=1} = d \sin \beta_{\text{fixed}} = m_2 \, \lambda_{m=2} \qquad \{4.17\}$$

or in a more generalized form, after rearranging, Equation {4.17} becomes:

$$\frac{m}{m+1} = \frac{\lambda_{m+1}}{\lambda_m} \qquad \{4.18\}$$

Regardless of the grating constant, for first and second order overlapping, this results in:

$$\lambda_{m=1} = 2 \, \lambda_{m=2}$$

which means that $\lambda 4000$ from order 2 overlaps exactly with $\lambda 8000$ from order 1 [82]. If we measure between $\lambda 4000$ and $\lambda 7500$ (silicon based CCD cameras!) the wavelength ratio $\lambda 7500/\lambda 4000$ is not greater than factor 2, so the first order spectrum will not be contaminated with wavelengths from order 2.

The physical meaning of the order overlap is that for a given incident angle α and grating spacing d, several wavelengths come together at a specific diffraction angle β. The mathematical equation {4.18} can be rewritten as:

$$\Delta\lambda = \lambda_m - \lambda_{m+1} = \frac{\lambda_{m+1}}{m} \qquad \{4.19\}$$

where $\Delta\lambda$ is defined as the free spectral range (FSR) and determines the wavelength range without overlap. The amount of overlapping orders depends on the grating spacing, however, second and third order, occurring higher orders and partial orders, will always overlap! Luminosity and resolution of a grating, on the other hand are independent of the order overlap, but depend on the width of the grating (see Equation {4.22}).

To eliminate the overlapping orders, adaptation to the optimal blaze angle and the use of special filters and cross dispersers, also called order separators are necessary. This is certainly the case with Echelle spectra, where the overlapping orders are desired (see Section 5.4). An overview of critical wavelength ranges where order overlap can occur is presented in Table 4.3. Fortunately, as has been demonstrated earlier, most silicon based CCD cameras used by amateurs are sensitive in the wavelength range $\lambda 4000$–$\lambda 7500$. Overlapping contamination must be avoided mostly, as can be seen from Table 4.3, when recording spectra in the wavelength ranges of the infrared and ultraviolet zones or in the higher orders of the visible wavelength range. As a general rule for spectroscopic recordings in the short (blue) wavelength side of the visible spectrum, second order is preferred. First order spectroscopic recordings are well suited for the long (red + near infrared) wavelength side.

4.3.2 Manufacturing Process and Performance Parameters

According to Geiger and Scheel [100], Milliet Dechales (1621–1678), produced a kind of metal and glass grating in transmission and reflection mode. Another historical note in honor of the American astronomer David Rittenhouse (1732–1796) who made a "precursor" of all gratings. This consisted of a bundle of some 50 hairs, strung between threaded screws made of brass wire, served as a primal grating of 3.23 cm^2 to demonstrate the diffraction of light. However, the different types of modern gratings are mostly manufactured following three major procedures.

4.3.2.1 Ruled Gratings

The first ruled grating was made by the German optician Joseph von Fraunhofer (1787–1826) in an optical workshop in Munich in 1813. He used a diamond translation machine to rule grooves on a metal substrate. Ruling of glass material was carried out later on by Friedrich A. Norbert (1806–1881) in 1850. The excellent quality of his gratings led to their use by the Swedish physicist A. J. Ångström (1814–1874) for his measurements of elements in the solar spectrum. In 1880 the American physicist Henry A. Rowland (1848–1901) made a grating of a large format and also introduced the first concave grating. Albert A. Michelson (1852–1931), another American physicist and known as the inventor of the Michelson interferometer, produced a ruling engine of a better quality. Based on this knowledge the modern ruling engines, known as the MIT B engines, were made by George R. Harrison (1898–1979), who was professor of physics at the Massachusetts Institute of Technology (MIT). The first real industrial production method, which dates back to 1947, concerns the mechanically high quality ruled gratings by the use of diamonds. For the first time high quality ruled gratings, highly comparable and with high perfection were produced. It was the result of the thorough work of the American astronomer Horace W. Babcock (1912–2003) at Mt Wilson and David A. Richardson at the Baush + Lomb company. Using a ruling engine a high quality and very expensive master grating is produced. In this process parallel grooves are ruled on a polished glass, fused silica or metal-based (copper) substrate.

4.3.2.2 Holographic Gratings

The second procedure is a photolithographic method, dating back to 1960. Holographic master gratings are manufactured by exposing photosensitive coated material with a holographic interference pattern, produced by intersecting laser beams, mostly Ar$^+$ ion laser at λ4880. The coated material used is photosensitive chalcogenide glass (ChG). The chalcogenide (ChG) coating consists of As_2Se_3, As_2S_3 or combinations of As-S-Se. Etching and reflective coating are carried out as post-processing techniques. The result is a holographic master grating with a sinusoidal profile.

4.3.2.3 Volume Phase Holographic Grating

The third type of grating is volume phase holographic grating (VPHG), which is available as transmission or reflection gratings. Between two glass or fused silica plates a layer of transparent medium functions as the transmission material. The transparent materials are dichromated gelatin, polymers, niobate crystals ($LiNbO_3$ and $KNbO_3$) or one of the sillenites: BGO ($Bi_{12}GeO_{20}$), BSO ($Bi_{12}SiO_{20}$) and TSO ($Bi_{12}TiO_{20}$), which are used as coating for the first substrate plate. The layer is exposed to a holographic fringe pattern in the same way as the production of the holographic grating. Consequently the exposed layer is transferred through a battery of liquid baths with appropriate chemicals. This generates a modulation in the refractive index in contrast to a conventional reflection grating where surface related diffraction occurs (Bragg diffraction). Finally the second substrate plate is glued as a protection, hence the advantage for cleaning of this type of gratings. It is important to note that only the original transmission or reflection VPHG grating is used.

A next step, except for the VPHG gratings, is the production of good quality and less expensive replicas, which are the commercially available forms of the different grating types. During this step, instead of the original master, a silicone rubber/epoxy submaster is used for the replication procedure. The three-dimensional structure of the master grating is transferred to the replica by pressing the (sub)master grating into a resin layer, which forms the coating of a glass. This way the replicated grating has a similar optical performance of the master grating, but is less expensive and is well adapted to large scale production. Properties of the three major types of diffraction gratings are presented in Table 4.4.

Table 4.4 Physical properties of three major types of diffraction gratings

Grating type	End-product	Diffraction efficiency	Stray or ghost light	Material sensitivity	Temperature/ humidity sensitivity
Ruled	Replica	High	Medium	Fragile	Low
Holographic	Replica	Medium	Low	Fragile	Low
VPHG	Original	High	Low	Not fragile	High

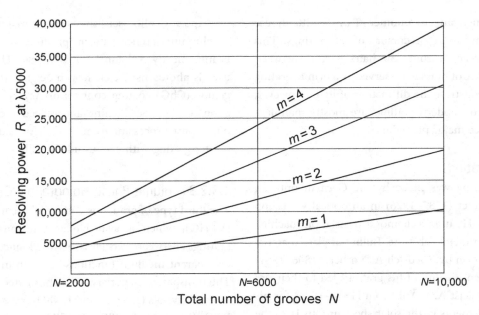

Figure 4.11 Evolution of the resolving power R at $\lambda 5000$, displayed as a function of the order m and total numbers of grooves N

4.3.3 Angular Dispersion

Indispensable parameters to estimate the final performance of the grating used in the spectrograph are the angular dispersion D, the reciprocal linear dispersion P, the resolving power $R_{grating}$ and the grating diffraction efficiency. The angular dispersion, D, gives the variation of the output angle β_m in the function of the wavelength as a result of constructive interference. With a constant angle of incidence α the angular dispersion of a grating can be calculated according to Equation {4.20}:

$$D = \frac{d\beta_m}{d\lambda} = \frac{m}{d\cos\beta_m} \qquad \{4.20\}$$

where $d\beta_m$ and $d\lambda$ represent the variation of the diffraction angle or angular separation and $d\lambda$ is the differential separation of the wavelength $(\lambda + \Delta\lambda)$, m is the order of diffraction and d the grating spacing. The angular dispersion is proportional to the order of diffraction m and inversely proportional to the spacing d. As a result the angular dispersion at the first order will be half of the one at second order, but, according to Figure 4.4, its luminosity or intensity will be still greater. The angular dispersion is expressed as degrees m^{-1}, arcmin or arcsec m^{-1} and radian m^{-1} (SI).

4.3.3.1 Reciprocal Linear Dispersion

Of course the spectrograph is more than just the grating alone, so the influence of the whole optical system (grating + collimator/objective + detector) comes in play here. The impact of the optical focal length can be estimated more conveniently by the so called reciprocal linear dispersion P. This equals the inverse linear dispersion D_{lin}, which is the product of the angular

dispersion D with the focal length f. As a result we get Equation {4.21} for the reciprocal linear dispersion P:

$$P = \frac{1}{DF} = \frac{1}{D_{lin}} = \frac{d\lambda}{dx} = \frac{d\cos\beta}{mf} \qquad \{4.21\}$$

where P is expressed in [Å mm^{-1}] or radians per nanometer [rad nm^{-1}]. The reciprocal linear dispersion P gives the maximum wavelength range for a specific length of the detector. Suppose we have a spectrograph with a P value of 30 Å mm^{-1}. It is coupled to a CCD camera, which can accept an output beam size of 15 mm in the direction of dispersion. The resulting spectral wavelength range obtained will then be 450 Å. So P is a very useful parameter in the development of a spectrograph. The value of the reciprocal linear dispersion increases with decreasing order of diffraction and vice versa.

4.3.3.2 Resolving Power

It is clear that the grating used must also be able to distinguish certain wavelengths by a smallest wavelength difference $\Delta\lambda$, which can be calculated by its resolving power R, here presented by Equation {4.22}, applicable only for planar gratings, which are most frequently used:

$$R_{grating} = \frac{\lambda}{\Delta\lambda} = mN \qquad \{4.22\}$$

where m is the order of diffraction and N is the total number of grooves (or lines) on the grating surface. A calculation of the theoretical maximum resolving power at $\lambda 5000$ of planar gratings with the same grating constant but increasing total numbers of rulings N is presented in Figure 4.11. It shows that

Table 4.5 Spacing, blaze angle and dispersion at λ5000 for five types of ruled gratings

Grating [Grooves per mm]	D [μm]	Blaze angle ($x°$ y')	Dispersion [nm mrad^{-1}]	Efficiency wavelength range [nm]
150	6.67	2° 8′	6.66	350–900
300	3.33	4° 18′	3.32	350–900
600	1.67	8° 37′	1.65	350–900
1200	0.83	17° 27′	0.80	350–900
1800	0.56	26° 44′	0.50	350–900

the R value of a planar grating generally grows with incremental order of diffraction and increasing total number of illuminated grooves. In practice, however, the theoretical maximum values of the resolving power are hardly reached and certainly never over the complete wavelength range. The influence of the other optical elements, like slit etc., is not considered here. Carefully manufactured gratings give values of R ending up to 90% of the theoretical maximum value.

4.3.4 Grating Diffraction Efficiency

Another important parameter of a diffraction grating is given by the diffraction efficiency. Two definitions are frequently used: the absolute and the relative efficiency. The absolute diffraction efficiency of a grating η_{abs} is given by Equation {4.23}:

$$\eta_{abs} = \frac{F_m}{F_0} \qquad \{4.23\}$$

where F_m represents the diffracted flux for a given order m and F_0 is the flux of the incident light. As η_{abs} represents a ratio it is expressed as a percentage and its graphical presentation is known as the absolute diffraction efficiency curve. The relative diffraction efficiency of a grating gives additional information on the reflectivity and is defined by Equation {4.24}:

$$\eta_{rel} = \frac{\eta_{abs}}{\rho_{coating}} \qquad \{4.24\}$$

where $\rho_{coating}$ represents the reflectance of coating, which is related to the reflectivity of the grating surface layer. It can be interpreted as the reflectance of a mirror with a similar coating as the grating. Important note here on the definitions: reflectivity is related to the material, reflectance to the sample with a specific coating. For the advanced DIY amateur

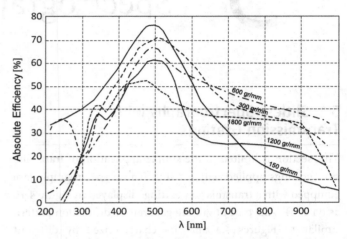

Figure 4.12 Absolute diffraction efficiency curves as simulated didactical illustration for 5 types of ruled gratings.

different manufacturer's catalogs are available to make an appropriate choice of the grating needed [78], [79], [80].

In contrast to the glass types used for prisms there is no generalized numbering or coding system used for gratings. Each manufacturer uses its own classification coding system based on the gratings parameters. Those parameters describe the type of grating, substrate and type of coating, number of grooves, blaze angle and wavelength and sometimes the original type of master grating used.

Important to note here that in manufacturer's catalogs about diffraction gratings, dispersion is defined as the reciprocal linear dispersion P and sometimes expressed in nanometer per milliradian abbreviated as [nm mrad^{-1}] or [nm mr^{-1}] instead of nm mm^{-1}.

Useful parameters of frequently used ruled diffraction gratings at a wavelength of 500 nm and optimized for use in the visible wavelength ranges are presented in Table 4.5 and their corresponding absolute diffraction efficiency curves in Figure 4.12.

5 Types and Function of Spectrographs

5.1 Slitless Spectrographs with Transmission Grating

This section is limited to the simple application of transmission gratings applied without slits. Those spectrographs are equipped with a transmission grating, displayed in Figure 4.10 as type (a). This type of grating generates a diffraction pattern similar to Figures 4.4 and 4.5 characterized by different orders. The starlight is incident on the back of the grating, generating the spectrum after transmission, hence their name. Incident light is mostly perpendicular to the back of the grating, therefore $\alpha = 0$, as shown in Figure 4.10(a). Zeroth order ($m = 0$) corresponds to direct transmission. Non-zeroth order diffraction ($m = -n$ or $+n$, i.e. order -2, -1, $+1$ and $+2$ in Figure 4.10) occurs with increasing wavelength and angles. For a good quality spectrograph a more advanced approach is reached when a specific spectral region is concentrated in a non-zeroth order, which is called a blazed grating. The design of a blazed grating is characterized by a typical triangular sawtooth profile.

The design with a linear optical axis makes a transmission grating very useful to a direct positioning in front of an objective lens or camera. Therefore it is a frequently used first type of spectrograph not only for the novice but later also for certain applications by the advanced amateur. Its easy installation and handiness enable a simple recording of a stellar spectrum which may be an incentive for purchasing. Anyway this simple application, without a slit in front of the dispersing element, is restricted mainly to point shaped or very small apparent light sources. Further, in contrast to a slit spectrograph, the achieved resolution depends strongly on the seeing conditions. In this way an absolute wavelength calibration with a light source is not feasible and therefore measurements with reasonable accuracy of radial velocities, not possible.

As already shown the resolving power is rising with increasing grooves per millimeter. However, if the density of grooves becomes higher the efficiency of the grating is reduced, certainly for the longer wavelengths as the finer spacing increases the diffraction angles. The more higher orders are generated, the more overlapping occurs. This limits also the resolving power of such types of gratings.

Transmission gratings can be screwed directly on the bottom side of an eyepiece (converging configuration) or in front of a lens of a CCD or DSLR camera (non-converging configuration). It is also possible to install those gratings in a filter wheel. In that case it will be occasionally necessary to adapt or change the holder of some gratings to make it fit in to the socket. For visual observation a dispersion lens is needed on top of the eyepiece to broaden the observed spectral stripe.

5.1.1 Available Transmission Gratings

Commercial gratings for amateur applications are available with 100, 200, 300 and 600 L mm^{-1}. They are first order blazed, which means that 75% of the energy of the light is gathered in the first order spectrum. Transmission gratings can be used with all types of telescopes and perform pretty well for all kinds of low resolution surveys. Different manufacturers offer gratings sold separately or available as a kit:

- The Rainbow Optics Star Spectroscope consists of a 100 L mm^{-1} grating with an additional cylindrical lens [62].

- The series of Paton Hawksley Education Ltd (PHEL) contains the well known Star Analyser (SA) in 100 L mm^{-1} and 200 L mm^{-1} versions [63].
- The "Blaze Gitter Spektroskop/Spektrograph" by Baader Planetarium consists of a 200 L mm^{-1} grating with an additional cylindrical lens (not available any more).
- The RS Spectroscope, manufactured by Rigel Systems, is a combination of a 600 L mm^{-1} grating and a dispersing cylindrical lens, both placed in a holder which can be mounted on the top of an eyepiece [64].

5.2 Slit Spectrographs with Reflection or Transmission Grating

In contrast to the slitless spectrographs the addition of a slit to the spectrograph concept makes a world of difference. The advantages of using a slit are the separation of the star of interest from neighboring stars, the elimination of stray light and the resulting higher resolving power. The slit width is an important parameter determining the throughput of the spectrograph, which is defined as the spread out in area and angle of the spectrograph's output.

A next step is the use of a reflection grating as the dispersive element of the spectrograph. As seen on Figure 4.10(b) the incident rays are on the same side of the grating as the reflected and diffracted rays. Similar to the transmission gratings is the distribution of light energy over the zeroth order or reflected angle and the different orders each containing a fraction of the light energy. To get more light energy in a certain order the aforementioned blaze technique is used to increase the amount of light energy in a chosen order. Blazed reflection gratings come in different versions as shown in Figure 4.10(c) depending on the angle position of the incident and diffracted rays. They are called nth order normal blazed or Littrow blazed. In the latter concept the angle of the incident ray α and the blaze angle Θ_B are the same.

5.2.1 The Classical Concept

The classical concept of a slit spectrograph is shown in Figure 5.1. The light is shown as it passes the slit and leaves the collimator as a parallel bundle of light rays. The reflection grating disperses the light, the spectrum of which is projected on the detector by a focusing objective.

5.2.2 The Littrow Design

Besides the classical concept another concept known as the Littrow configuration is presented in Figure 5.2. It combines the collimator and focuser in one and the same optical device. The main advantage of the Littrow configuration is the possibility to manufacture a linear design inside the spectrograph. As a minor drawback the spectral lines within the recorded stripe are displayed here significantly curved.

5.2.3 Configurations and Options for the Development of Slit Spectrographs

The configurations of Figures 5.1 and 5.2 are frequently used to develop prototypes of amateur spectrographs. Additionally, the collimator optical performance is very often adapted for the frequently used telescope optics by amateurs, being $f/10$-types. The most commonly used configuration is the small angle Littrow geometry. Additionally, it is well suited for mass production and the price setting is affordable for the amateur.

It is also clear that efforts have to be made to control the total weight of the spectrograph, preferably in a design as

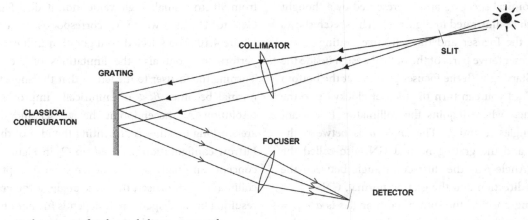

Figure 5.1 The optical concept of a classical slit spectrograph

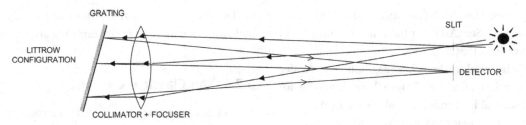

Figure 5.2 The optical concept of a Littrow spectrograph

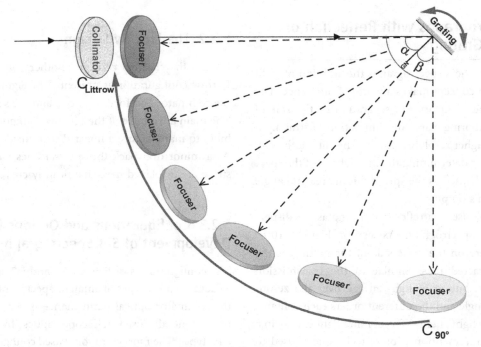

Figure 5.3 Illustration of possible spectrograph designs with planar reflective grating

compact as possible, as it must be mounted on the telescope together with guiding and recording camera devices.

A selection of configurations useful for the development of such an astronomical spectrograph is presented by a thought experiment such as illustrated in Figure 5.3. The sketch shows the collimator, the focuser and the reflective grating as the three main representative parts of the optical design inside the spectrograph. Starting with the focuser position at the bottom right corner ($C_{90°}$) you can turn the focuser clockwise to the left upper corner, where it joins the collimator. Important here are the angles α and β. The angle α is between the incoming rays and the grating normal GN, also called the grating angle. Angle β is the diffraction angle between the diffracted light direction and the grating normal, GN.

At the starting position the sum of both angles is $\alpha + \beta = 90°$. Depending on the chosen or desired geometry the value

of $\alpha \pm \beta$ is a typical parameter of the spectrograph. As a result this total angle between the incident rays at the collimator and the diffracted rays collected by the focuser can be fixed going from 90° to a small angle value indicated in Figure 5.3 from $C_{90°}$ to $C_{Littrow}$, where C corresponds to type (c) from Figure 4.10. The selected configuration influences the overall performance but also the limitations of the spectrograph. Turning the focuser to the left so that the angle to the grating normal becomes $\beta < \alpha$, empirically improves the spectral resolution. At the endpoint the focuser and the collimator are working together representing the idea of the small angle Littrow configuration ($\lambda_b = 2d \sin \Theta_b$ in Figure 4.10). Which concept can finally be used depends on the optical design of collimator and focuser, the blaze angle of the grating and the resulting linear dispersion. It depends further on the intended area of astronomical applications. The resulting resolving power R of each configuration can be calculated by substituting the order of diffraction m in the grating equation {4.16} into Equation {4.22}, which results in Equation {5.1}:

$$R = \frac{W(\sin \alpha + \sin \beta)}{\lambda} \tag{5.1}$$

where W is the ruled width of the planar grating. The theoretical maximum value of R is reached when $\alpha = \beta = 90°$ which results in $R_{max} = 2W/\lambda$. A more realistic and attainable value is reached with ~70% of the theoretical value, which results in a value of ~$4W/3\lambda$. Calculations of R suppose the grating is fully illuminated, which in reality is not the case, depending on the width of the slit used, which contributes to the overall resolving power of the spectrograph.

5.2.4 Anamorphic Magnification

The combination of the collimator–grating–focuser–camera position plays a key role in the final performance of the spectrograph. Depending on the focal ratios of collimator and focusing objective respectively with rotation and tilt of the grating and variable slit width, the resulting image of the spectrum will be influenced. This is indicated by the term: anamorphic magnification. The phenomenon was first reported by Ira S. Bowen in 1952. Later it was called the anamorphic magnification by Chaffee and Schroeder in 1976 and was brought again under attention by F. Schweizer in 1979. This effect, generated here by changing the inclination of the incoming light on a grating, is demonstrated in Figure 5.4.

On Figure 5.4 the angle of incidence α is significantly greater than the angle of diffraction β. The effect of the anamorphic magnification is clearly visible by the difference between d and D. The anamorphic magnification r, caused by the grating, can be calculated by differentiating the grating equation {4.16} which results in Equation {5.2}:

$$r = \frac{|d\beta|}{|d\alpha|} = \frac{\cos \alpha}{\cos \beta} = \frac{d}{D} \tag{5.2}$$

In case $\alpha = \beta$ (Littrow configuration), for example, an angle of 25° results in $r = 0.906/0.906 = 1$, which means there is no anamorphic magnification. Where $\alpha < \beta$ this results in $r > 1$,

a situation called magnification. In this situation the projected wavelength range per mm slit image increases, which negatively influences the resolution as less spectral information is transferred per unit of slit image. The opposite, $\beta < \alpha$, gives $r < 1$ and is defined as demagnification. The latter is the most favorable situation to improve the spectral resolution as more spectral information is transferred per unit of slit image! However, $r < 1$ on the contrary decreases the efficiency. Therefore manufacturers have to make a balance between anamorphic magnification and overall efficiency.

Besides the effect of the grating, the ratio of the focal lengths of the collimator and the camera contributes also to the resulting projected slit width w, which can then be calculated according to Equation {5.3}:

$$w = r \frac{f_{\text{camera}}}{f_{\text{collimator}}} W_{\text{slit}} \tag{5.3}$$

where W_{slit} is the original slit width, r is the anamorphic magnification, f_{camera} and $f_{\text{collimator}}$ represent the respective focal lengths, and w gives the final dimensions of the so called "spectral stripe" registered by the camera and which is typical for the spectrographic configuration used.

The influence of the collimator–camera angle Φ defined by $\alpha - \beta$ and the grating tilt t on the anamorphic magnification represented by Equation {5.2} can be written in another form as Equation {5.4}:

$$r = \frac{\cos \left(t + \frac{\Phi}{2} \right)}{\cos \left(t - \frac{\Phi}{2} \right)} \tag{5.4}$$

where the grating tilt t is defined as the angle between the bisector BS (of the incident ray coming from the collimator and the diffracted ray going in the direction of the camera) and the grating normal GN as illustrated in Figure 5.5.

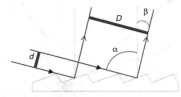

Figure 5.4 Anamorphic magnification by a grating at large angle of incidence

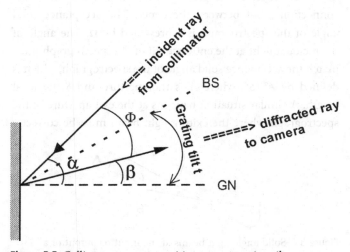

Figure 5.5 Collimator–camera position versus grating tilt

If, according to Equation {5.4}, the tilt of the grating increases, it becomes possible to make the slit width larger to a certain limit without losing resolving power. This way more light gets in the spectrograph and recording times become smaller. For the different angle positions of Figure 5.3 between collimator and focuser relative to the grating normal the corresponding anamorphic magnifications can be calculated. In case of spectrographs based on reflective gratings the calculated r values can go up to 1.5 and even higher.

5.2.5 Spectrograph Throughput and Etendue

One of the important parameters of a spectrograph is the "total efficiency," defined by the product of light throughput and resolving power $\Theta \times R$ [5]. Finally the overall throughput or etendue Θ of a spectrograph is defined as the passing light intensity in W or photons/s and can be calculated by Equation {5.5}, which represents also the light-gathering power of a spectrograph. It remains constant between all optical elements, influencing the whole light path.

$$\Theta \approx A\Omega \qquad \{5.5\}$$

Here, A is the area of the minimum aperture, defined by the product of the width and the length of the slit and Ω is the solid angle accepted by the spectrograph entrance aperture. It is the ratio of an observed or projected surface area to the square of the radius inside an imagined ambient sphere. The source, emitting light which passes through the slit of the spectrograph, has a certain input or output area respectively. This generates a solid angle subtended at the input of the spectrograph as shown in Figure 5.6. The sketch shows two cones. The left light cone comes from the collimator and the right one goes in the direction of the grating, after passing the slit at the connection point between the cones. The acceptance solid angle of the spectrograph is represented by Ω. The angle of the incident light at the entrance slit of the spectrograph must match the acceptance solid angle of the spectrograph, which is defined by A^2/S^2, where A is the aperture and S the focal length. A similar situation happens at the exit aperture of the spectrograph where the exiting light beam must be correctly

positioned in front of the detecting camera. Although expressed in sr (steradian), Ω is dimensionless. Finally, the calculated etendue value Θ is expressed in m^2 sr (or cm^2 sr). It is clear that the necessary compromises have to be made by manufacturers in the design of their spectrographs. The pros and the cons of each concept will have to be balanced in respect to the application the spectrograph is made for.

5.2.5.1 Rowland Circle

Commercially available spectrographs vary from the 90° design (e.g. DADOS) to the Littrow based designs (e.g. Spectra L200, Lhires III). Besides the use of a planar reflective grating some manufacturers have developed an optical design based on a curved reflective grating (e.g. SX Spectrograph). This type of optical design is known as the Rowland circle. Therein the positions of the incoming light from the slit and the diffracted light recording point are positioned on a circle following the radius curvature of the concave grating as shown in Figure 5.7.

The pure Rowland circle based spectrograph is not commonly used in astronomy, but with some modifications, especially concerning the flat instead of the curved appearance of the generated spectrum and the optimization of the stigmatic points, this type of spectrograph can also be used for the recording of astronomical spectra. Therefore a special type of grating is used: the curved toroidal grating, which has different equatorial and meridional curvatures to correct the astigmatism frequently observed with concave grating designs. The compact design of such type of spectrograph is an extra advantage.

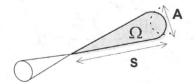

Figure 5.6 Solid angle light beams at the input or output of a spectrograph

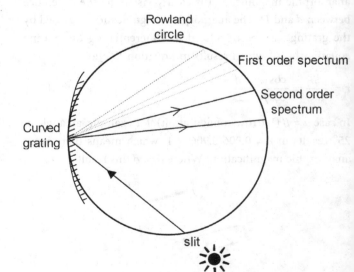

Figure 5.7 Spectrograph based on the Rowland circle design

5.3 Commercial Slit Spectrographs for Amateur Applications

Table 5.1 shows in alphabetical order an overview of present available amateur astronomical slit spectrographs with reflection or transmission gratings. All spectrographs are operating from approximately the border of the ultraviolet range over the visible range to the near infrared.

5.3.1 ALPY, Shelyak Instruments

ALPY is a slit spectrograph manufactured by Shelyak Instruments [59]. It is equipped with a so called grism, a combination of a transmission grating (600 L mm^{-1}) and a prism. The removable standard slit includes multiple positions – 25, 50, 100 or 300 μm slits, a 25 μm hole and a clear position with a 3 mm hole for a slitless mode. This highly compact design reaches a resolution of $R \sim 600$. This multipurpose device with a very low f-number of f/4, typical for large Dobsonian telescopes, is optimized to record very faint objects like extragalactic supernovae.

5.3.2 DADOS, Baader Planetarium

DADOS is a conventional slit spectrograph manufactured by the German Baader Planetarium [58]. It can be operated with three different gratings 200, 900 or 1200 L mm^{-1}, corresponding to resolutions at $\sim\lambda6000$ of $R \approx 900$, 4000 and 5000. It has been shown that even the application of an 1800 L mm^{-1} grating is possible, however, with a significantly reduced observable wavelength range of $\lambda\lambda3900$–5500 (excluding the Hα line) and reaching a resolution of just $R \sim 6000$. As a result of the 90° design the use of higher resolving types of grating is very limited. The available slit widths are 25, 35 and 50 μm. The device is connected either directly to the telescope adapter or via an interconnected flip mirror. DADOS is optimized for f/10 telescope optics.

5.3.3 Lhires III, Shelyak Instruments

The Lhires III is a Littrow type spectrograph, manufactured by Shelyak Instruments [59]. It is equipped with a standard 2400 L mm^{-1} reflective grating reaching resolving power of $R = 18,000$ at $\lambda6563$. Other gratings are available going from 150 L mm^{-1} to 2400 L mm^{-1}. The standard four-postion slit of 15, 19, 23 and 35 μm can be extended with a 50, 75, 100 μm slit and a 19 μm hole kit. Additionally a spectrophotometry kit is also available. The spectrograph contains an internal argon–neon–helium calibration lamp and a tungsten lamp to record a flat field. The device is optimized for f/10 telescope optics.

5.3.4 LISA, Shelyak Instruments

The Long slit Intermediate resolution Spectrograph for Astronomy (LISA) is a low resolution high luminosity spectrograph, equipped with a reflective slit specially designed for faint and extended astronomical objects. Therefore it can be used for the study of nebulae, novae, galaxies, quasars, variable stars and comets, as well as for a quick recording of low resolution spectra of stars and nebulae. The basic spectrograph module consists of a guiding compartment with reflective slit (15, 19, 23, 35 μm) and optical compartment with a 300 L mm^{-1} grating and is optimized for f/5 telescope optics. Resolving power is between $R = 600$ and $R = 1000$. The basic module can be upgraded with a calibration module and a near-IR feature and the slit can be extended to 50, 75, 100 μm and photometric type.

5.3.5 Minispec, Astro Spectroscopy Instruments

The Minispec is a classical optical design based spectrograph manufactured by the German company Astro Spectroscopy Instruments EU. The basic version comes as a slitless device, which can be upgraded later with a slit and calibration unit. Grating options range from 300 L mm^{-1} to 1800 L mm^{-1} inclusive versions adapted for NIR. The spectrograph can be adapted for f/5 or f/10 telescope optics [65].

5.3.6 Spectra L200, JTW Astronomy

The Spectra L200 spectrograph is a Littrow based instrument, manufactured by the Dutch company JTW Astronomy and following the concept created by Ken Harrison. Standard equipment contains a 600 L mm^{-1} reflective grating, which can be upgraded to 1800 L mm^{-1}. The reflective slit is equipped with multi-indexed wheel to make a choice between nine gaps (20, 25, 30, 35, 40, 45, 50, 75 and 100 μm) and three pinholes (25, 50 and100 μm). The Spectra L200 is optimized for telescope optics with focal ratios in the range f/7 to f/10, depending on the grating used [66] [67].

5.3.7 Starlight Xpress SX, Starlight Xpress

The SX spectrograph is a compact modified Rowland circle based spectrograph, manufactured by the UK company

Table 5.1 Commercial astronomical spectrograph types

Spectrograph	Exterior View	Design $\alpha + \beta$	Grating Type	Grating L mm^{-1}	R_{max}	Collimator Focal Ratio	Weight [kg]
ALPY		Linear	Grism	600	600	$f/4$	0.12
DADOS		90°	Planar	200–1200	6000	$f/10$	0.85
Lhires III		Littrow	Planar	150–2400	22,000	$f/8$	1.7
LISA		Classical	Planar	300	1000	$f/5$	1.40
Minispec		Classical	Planar	300–1800	5000	$f/4$	0.5

Table 5.1 (*cont.*)

Spectrograph	Exterior View	Design $\alpha + \beta$	Grating Type	Grating L mm^{-1}	R_{max}	Collimator Focal Ratio	Weight [kg]
Spectra L200		Littrow	Planar	150–1800	9000	$f/7$	1.2
Starlight Xpress SX		Modified Rowland Circle	Concave Toroidal	550	2000	$f/5$	1.17

Starlight Xpress. The grating has 550 L mm^{-1} and the maximum attainable resolving power is $R = 2000$ at $\lambda 5000$. A six position slit wheel varies the slit from 20 to 300 µm, inclusive a hole of 3×3 mm. To make it a complete "plug and measure" device an internal battery operated calibration lamp and guiding camera are already installed. The spectrograph accepts fast telescope optics from $f/3.5$, but optimal use is from $f/5$ and above [68].

5.4 Echelle Spectrograph

5.4.1 Overview

The technological evolution at the time of Henry Rowland inspired the American physicist Robert Williams Wood (1868–1955) together with his friend Edward Shane of the Lick Observatory, to make all kinds of gratings, from the conventional types, called the echelettes to combinations of gratings in mosaic forms. Simultaneously the number of rulings was also lowered which gave birth to a new form of grating: the echelle, from the French word echelle which means a ladder. The real breakthrough happened with the work of George R. Harrison at MIT. Typical for the echelle grating is the wide spacing of the grooves, from 20 L mm^{-1} to roughly 300 L mm^{-1} and the high blaze angles, mostly from 63° to 78°. Numerous overlapping spectral stripes with very high orders are generated in this way. To get access to these

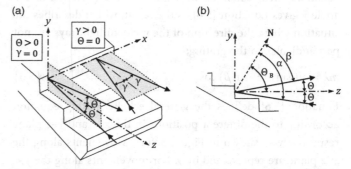

Figure 5.8 Off-axis angle positions of the light pathway on echelle gratings. 3D view and longitudinal section by M. Huwiler, schematically after [81] [83].

highly resolved spectra they must first be spread in the vertical direction by use of a so called cross disperser, which can be a prism, a grating or a grism. However, the resulting high dispersion of this principle enables the manufacture of high resolution spectrographs. Moreover, with just one single recording nearly the whole optical range can be covered! A general illustration of the light pathway of an echelle grating is given in Figure 4.10(d). Figure 5.8 shows additionally a more detailed sketch of the echelle grating optics where α represents the incident angle, β the angle of diffraction, γ the angle between the incident ray and the y–z plane, and Θ the angle between the incident ray and the x–z plane.

5.4.2 Basic Designs of Echelle Spectrographs

According to Figure 5.8 and considering the off-axis angle positions, most of the echelle spectrographs can be subdivided into two main designs.

Case 1 $\Theta > 0$, $\gamma = 0$. The main features of this design are nearly straight running orders, enabling in a good approximation a simple processing and calibration of individual orders. The achievable resolution is somewhat higher compared with case 2. Typical examples for this concept in the amateur sectors are Baches [58] and SQUES [60].

Case 2 $\gamma > 0$, $\Theta = 0$. The main features of this "quasi Littrow" design are rainbow-like appearance, more or less strongly curved orders, requiring an overall processing of the recorded echellogram with specific software. The optical efficiency of this setup is some percent higher compared with case 1. A typical example for this concept is eShel [59].

According to the work of D. J. Schroeder and R. L. Hillard [81] only two light pathways, as indicated by configurations D_1 and D_2 in Figure 4.10, make sense in practice, represented by $\alpha > \beta$, which means $\Theta > 0$ and $\alpha = \beta = \Theta_B$ which means $\Theta = 0$.

Starting for Case 1, $\Theta > 0$, $\gamma = 0$, with the general grating equation {4.16} extended with the different positions of the angle γ gives Equation {5.6} as the result, which describes the situation where the direction of the incident light rays are not perpendicular to the grating;

$$m\lambda = d(\sin\alpha \pm \sin\beta)\cos\gamma \qquad \{5.6\}$$

Equation {5.6} defines the general echelle grating equation according to the different positions in the x–z and y–z plane respectively as shown in Figure 5.8(a). Movements along the x–z plane are represented by γ. For movements along the y–z plane α and β equal $\Theta_B + \theta$ and $\Theta_B - \theta$ respectively, as shown in Figure 5.8(b). The general equation {4.20} describing the grating angular dispersion is then adapted for the echelle situation and transformed into Equation {5.7};

$$\frac{d\beta}{d\lambda} = \frac{m}{d\cos\beta\cos\gamma} \qquad \{5.7\}$$

Solving Equation {5.6} for m and combining it with Equation {5.7} results in Equation {5.8}:

$$\frac{d\beta}{d\lambda} = \frac{\sin\alpha + \sin\beta}{\lambda\cos\beta} \qquad \{5.8\}$$

It gives the angular dispersion for configuration in Figure 5.8 where $\theta > 0$.

The other configuration in Figure 5.8 where $\gamma > 0$, $\Theta = 0$ known as the "quasi Littrow" mode, where both incident and diffracted angles quasi-equal the blaze angle θ_B. In that case the corresponding angular dispersion, given by Equation {5.8}, can be written as in Equation {5.9}:

$$\frac{d\beta}{d\lambda} = \frac{2\tan\theta_B}{\lambda} \qquad \{5.9\}$$

This equation forms the basis for a classification or reference system for echelle gratings. The angular dispersion increases with increasing blaze angle. The tangent of the blaze angle corresponds to an R number. Important note here: This echelle reference number R must not be confused with R for resolving power.

An overview of the most frequently used echelle grating R types is presented in Table 5.2.

As can be seen from Figure 4.10 an important difference can be noticed upon the illuminated sides of the grooves on the grating used. The conventional gratings indicated by (a), (b) and (c) are used by their long side of the grooves. On the contrary, the echelle gratings are used by their shortest side. Additionally, using a large blaze angle increases the dispersion impressively together with the generated multiple higher orders. As a result a much higher resolving power can be achieved. The only downsides are the overlapping orders. Therefore cross dispersers are used to spread the light perpendicular to the echelle dispersion direction in order to sort the different orders. Factors influencing the separation of the different orders are the angular dispersion of the optical device, used as cross disperser (prism, grating or grism), the focal length of the recording camera lens and the free spectral range, defined by Equation {4.19}. The result is a two dimensional (2D) spectrum or echellogram, as illustrated in Figure 6.6, with order numbers increasing downwards and wavelength increasing from left to right. A visible side effect that occurs with the use of the cross disperser is the tilted and curved lines, but this effect poses no problems for a correct software reduction later on.

The theoretical resolving power of an echelle grating can be calculated by combining Equation {4.22} with {5.6}, assuming

Table 5.2 *R*-number classification of echelle gratings

Blaze Angle	R
63.4°	2
71.5°	3
76°	4
79°	5

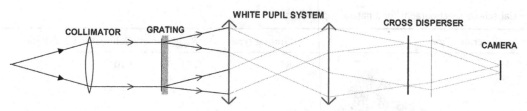

Figure 5.9 White pupil optical design

the off axis angle γ is very small to approximately zero, which finally results in Equation {5.10}:

$$R = \frac{2W}{\lambda} \sin \theta_B \qquad \{5.10\}$$

It is similar to Equation {5.1} for the conventional grating, but emphasizes here the more important proportionality of the resolving power R to the blaze angle θ_B which is also the great advantage and strength of the echelle grating.

The efficiency of an echelle spectrograph can be significantly increased by the optical design concept of the so called "white pupil" or Baranne's mounting. This effect is mainly achieved by the reduction of stray light and the better match between resolution and pixel size, as the beam size of the white pupil can be adapted to the camera aperture. This makes sense for very large spectrographs with accordingly large sized echelle gratings. Therefore this design, based on the work of the French optical engineer André Baranne, is now frequently used on numerous professional, high resolution echelle spectrographs as ELODIE, UVES, HRS and FEROS. The success of this concept will inspire spectrograph designers to develop double and even multiple white pupil designs for the future.

The concept of the white pupil can be interpreted as a re-imaging of the grating pupil on the cross disperser. Therefore additional optics, which can be a lens or a mirror system, is installed between the echelle grating and the cross disperser. Because of this additional optics the dispersed light beams, coming from the grating, overlap each other at this generated intermediate image, appearing as a white light, which explains also the origin of the name "white pupil." This optical design is shown in Figure 5.9.

5.5 Commercial Echelle Spectrographs for Amateur Applications

5.5.1 BACHES Spectrograph, Baader Planetarium

The **Ba**sic **Ech**elle **S**pectrograph, BACHES, is an echelle grating based spectrograph for direct coupling on the telescope [58]. It is optimized for 8″ to 24″ *f*/10 telescope optics

and reaches an average resolving power of $R = 18{,}000$. BACHES is provided with two selectable slits: 130 μm × 25 μm and 130 μm × 50 μm. The spectrograph can be extended with a calibration unit, mounted on the spectrograph's housing or as a remote calibration unit controlled via a web interface.

5.5.2 eShel, Shelyak Instruments

This optical fiber fed echelle spectrograph comes in two versions [59]. The standard eShel version reaches a resolving power of $R > 10{,}000$. For calibration a separate Th-Ar unit with 230 V power supply is available. Also an eShel+ version is under construction striving for a resolution of $R \approx 30{,}000$. It is intended that an optical bench added to the configuration will optimize the use of the white pupil optical design. Both systems use a fiber-based interface between the telescope and the spectrograph/calibration unit. This enables the spectrograph to be mounted remotely in an acclimatized box. This remote concept, connected by a 50 μm optical fiber, is optimized for use with larger telescopes and for high precision measurement of radial velocities.

5.5.3 SQUES, Eagleowloptics

The "**S**wiss **Qu**ality **E**chelle **S**pectrograph" SQUES [60] is Swiss made and designed for direct coupling on ~5″ to 40″ telescopes, optimized for focal ratios of f/9 to f/12. It reaches a maximum resolving power of ≈ 24,000 and is equipped with an adjustable slit width in the range of ~15 to 85 μm. This allows a fine tuning to the telescopes spot size, the pixel raster of the camera and the seeing conditions (Figure 6.3 and Table 6.1). The delivery includes a small Ar-, Ne-, He-calibration unit, mounted on the spectrographs housing, supplied with 12 V DC. The according glass fiber socket allows the calibration with any other light source. Finally as an option, an interface for a glass fiber connection between spectrograph and telescope is available. An overview of the present commercial available echelle spectrographs is presented in Table 5.3.

Table 5.3 Commercial echelle spectrographs for amateur applications

Spectrograph	Exterior View	Grating type	R_{max}	Collimator Focal Ratio	Weight (kg)
BACHES		R2	20,000	$f/10$	1.35
eShel		R2	10,000	$f/5$	N/A
eShel+		R2	30,000	$f/6$	N/A
SQUES		R2	24,000	$f/10$	1.02

5.6 Czerny–Turner Spectrograph

5.6.1 Overview

A design based on the use of two mirrors is called the Czerny–Turner spectrograph. It was presented by the German physicist Marianus Czerny (1896–1985) and the American physicist Arthur Francis Turner (1907–1996) in 1930 [10], and was inspired by the English astronomer and chemist Sir William de Wiveleslie Abney (1843–1920), who experimented in 1885 with a two mirror configuration. The configuration of this type of spectrograph is presented in Figure 5.10 where S1 represents the entrance slit, PG the planar grating and M1 and M2, the collimating and camera mirror respectively. At point D a camera or second slit can be installed. In case a slit is installed the spectrograph is in monochromatic mode, this is frequently used in laboratory conditions. In spectrograph mode a camera is installed to directly register the spectrum. Additionally, in solar astronomy the Czerny–Turner spectrograph is frequently used in the double pass mode. In this mode the beam coming from M2, instead of exiting the spectrograph, is deviated by a small mirror–slit–mirror optical arrangement and passes

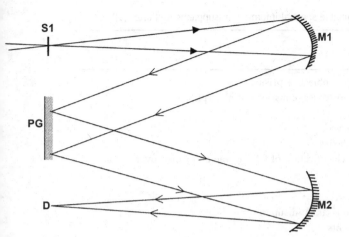

Figure 5.10 Light pathway in Czerny–Turner spectrographs

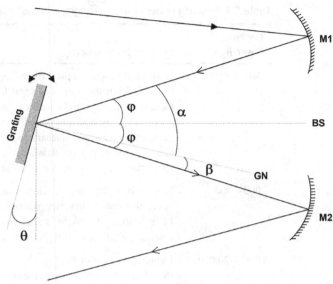

Figure 5.11 Angle Positions and Variables

the grating again. The advantage of this double pass is to make corrections for interferences of scattered light in the spectrograph, which is critical for precise measurements in the short wavelength bands. An example is the McMath–Pierce solar telescope of Kitt Peak National Observatory.

A typical feature of the Czerny–Turner design is a very high resolution of typically $R > 50,000$ up to $R > 1,000,000$, displaying just a very small bandwidth of a few ångstroms.

Apart from the planar grating, PG, all parts are fixed. Incident light passing through the entrance slit, S1, falls on the curved mirror, M1, where it is collimated. The parallel bundled light reaches then the plane grating, PG, where it undergoes diffraction and makes its way to the second curved mirror M2, which functions as a focusing tool, positioning the rays to pass the exit slit S2 to finally arrive at the camera. The mirrors M1 and M2 are concave spherical mirrors and may have different curvatures. This way possible aberrations generated at M1 are corrected by M2. The design of such a spectrograph is also known as the "W" configuration. The general form of the grating equation can be adapted to a more convenient form corresponding to the optics of the Czerny–Turner spectrograph.

The angle positions and the variables are presented in the sketch of Figure 5.11 where α and β represent the incident and diffracted angles relative to the grating normal GN. Here φ represents the angle between incident and diffracted rays and the bisector BS of both rays. The angle θ gives the rotation position angle of the grating and also equals the angle between GN and BS. For this configuration $\sin \alpha = \sin (\varphi - \theta)$ and $\sin \beta = \sin (\varphi + \theta)$ applies. Substituting in the general grating Equation {4.16} results in:

$$m \lambda = 2d(\sin \theta \cos \varphi) \qquad \{5.11\}$$

According to Equation {5.11} the rotation of the grating ($\sin \theta$) is proportional to λ.

5.6.2 Application of Czerny–Turner Spectrographs for Amateurs

In the amateur sector large Czerny–Turner spectrographs are mainly applied for high resolution solar spectroscopy and spectro-heliography. Frequently converted devices are used which originate from chemical or physical laboratories. Commercial spectrographs, designated for the amateur market are not available yet, although prototypes are being developed.

5.7 Spectral Information and Required Resolution

Each of the discussed spectrographs has its own characteristics and applications. Depending on which astronomical object will be studied the resolving power of the spectrograph plays a key role in the choice. An important turning point to be able to analyze line profiles and Doppler shifts is a value of $R = 10,000$. Such details are in high demand by professional astronomers. Therefore high resolution spectrographs are essential for pro-am collaborative projects. Table 5.4 represents a rough overview of the type of spectral information that can be found in the recorded spectrum with increasing resolving power R [9].

Table 5.4 Resolving power of the spectrograph versus obtainable spectral information (supplemented after [9])

Resolving Power R	Spectral Information
150–2000	Wide field spectroscopic surveys e.g. by objective prisms General determination of spectral features e.g. emissions or absorptions General stellar spectral classification Spectral energy distribution (SED curves) Redshift of very faint quasars and galaxies General stellar spectral classification classification of faint novae and supernovae Excitation class of emission nebulae
2000–9000	Details for stellar spectral classification Spectral details of brighter galaxies, quasars and supernovae Identification of molecules and elements Analysis of element abundance and metallicity
10,000–20,000	Detailed analysis of line profiles e.g. for disks of Be stars Rotation velocities of planets and stars Stellar temperature, analysis of particular sensitive lines
20,000–50,000	Detailed analysis of line profiles (e.g. for Wilson–Bappu effect) High precision measurements of radial velocities Analysis of solar and stellar magnetic fields by Zeeman effect
50,000–100,000	For very large telescopes: Doppler analysis and mapping of winds, circumstellar and proto-planetary disks, flares, interstellar medium
> 100,000	For very large telescopes: atmospheric structures, thermal broadening, analysis of interstellar lines, chemical composition of exoplanetary atmospheres. For solar telescopes: detailed analysis of solar surface and granulation

However, it makes no sense in any case to strive for the highest possible resolution. As a general rule: The higher the resolving power the longer the required exposure time. Thus for high resolution analysis of faint objects accordingly large professional telescopes are needed. Even with such equipment, for extremely distant and accordingly faint quasars and supernovae, low resolution spectrographs are applied. Further, for rough stellar classification lowly resolved broadband spectra are preferred, displaying at a glance all relevant, spectral features.

CHAPTER

6 Recording of the Spectra

6.1 Visual Observation of Spectra

Despite very early attempts to record the spectra photographically in the first half of the nineteenth century spectra have been mainly observed visually and documented by drawings. Famous examples have been preserved mainly by Joseph Fraunhofer and Father Angelo Secchi. Various examples can be found in the *Spectral Atlas* [1]. But still today most amateurs take their first steps by observing visually, frequently using DIY equipment like CD-ROMs and the like. So a short excursus on the functions of our own "sensor" may be meaningful here. It is also interesting to look for parallels but also differences to electronic image sensors.

In the outermost layer of the retina of our eyes there are two types of photoreceptor cells, called rods and cones. A schematic view is presented in the sketch in Figure 6.1.

6.1.1 Scotopic or Nighttime Vision

There are about 120 million rods which are sensitive to light and contain the pigment-containing protein rhodopsin,

which has its maximum absorption at 4980 Å or 498 nm. They give us the so called scotopic vision, which means the ability to see under reduced light conditions or dark skies. This ability is important for the dark adaptation of our eyes when doing visual deep sky observations. The rods are not sensitive to colors, which explains the sense of using red lights during astronomical observations. The longer wavelength of the red light does not interfere with the sensitivity of the rods. This effect is known as the Purkinje effect: The sensitivity of our eyes shifts toward shorter wavelengths under dark skies. Therefore when observing a nebula through the eyepiece of a telescope we only can see a hazy gray-green spot.

6.1.2 Photopic or Daytime Vision

The 6 million cones are sensitive to colors and contain the protein-pigment photopsin, which mostly appears as three types with maximum absorptions at 420, 533 and 564 nm. Sensitivity to three colors or possessing three channels for

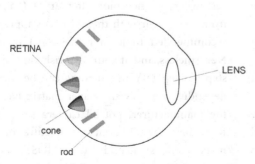

HUMAN EYE

λ_{max}	Wavelength range of absorption
420 nm	370 – 530 nm
533 nm	400 – 650 nm
564 nm	400 – 680 nm
498 nm	400 – 600 nm

Figure 6.1 Spectral characteristics of photoreceptors in the human eye

conveying color information is called trichromacy. Having four such channels is called tetrachromacy. Other species such as birds, butterflies and bees can have even more types with maximum absorption up to the ultraviolet wavelength range (370 nm). The cones that give us color or photopic vision are positioned in the more central region of the retina. The peak sensitivity of our eyes in daylight lies in the yellow-green wavelength range.

When we look at a light source we see a composed color impression which reflects the overall spectral light distribution of all present wavelengths visible to the photoreceptors in our eyes. This creates our visual color perception, though it is not possible to distinguish the individual spectral characteristics. For the latter we need a spectrograph to observe the different spectral lines of a light source such as gas discharge lamps like compact fluorescence lamps (CFL) or energy saving lamps (ESL) where the light appears as white to our eyes.

6.2 Recording of Spectra with Electronic Image Sensors

Today for a scientifically relevant activity a photographic recording of the spectra is indispensible. The different types of cameras and their operating principles are extensively described in numerous publications about astrophotography. Next we make some supplementary remarks regarding spectroscopically relevant peculiarities.

For the professional, as well as advanced amateur spectroscopy, in most cases dedicated, cooled astronomical monochrome CCD cameras are applied. They best meet the most important requirements – a combination of low noise and high sensitivity. Also important is the binning function, enabling us to combine a cluster of pixels, mostly 2×2 or 3×3, into single blocks with significantly increased sensitivity. This way the recording of very faint objects is enabled and, as outlined in Section 6.3.6, the pixel grid of the camera can be optimized and adapted to the slit width and the optical design of the spectrograph. Also the size of the sensor must fit to the given camera lens. If a new camera is to be acquired, it is worthwhile, possibly with test flats, first to check whether it produces interference fringes to a disturbing extent (Section 6.3.7). Figure 6.2 shows the quantum efficiency (QE) of the Sony chip ICX 285, which represents a typical example for the characteristics of most sensors applied in today's astronomical cameras. Compared to the eye the sensitivity here ranges much further in to the UV and IR. In contrast to astrophotography, in spectroscopy no

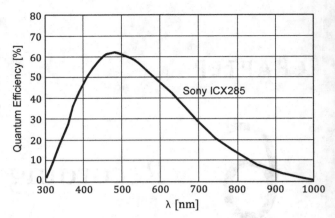

Figure 6.2 Wavelength dependent quantum efficiency of the Sony chip ICX285

blocking filters are applied and the whole wavelength domain is used as far as possible.

Quantum efficiency is defined as the wavelength dependent efficiency with which the sensor converts incident photons into electrons.

Due to the Bayer matrix, color cameras have a reduced sensitivity. Also the mostly useless color information, generated here by the 2×2 pixel blocks, may be a hindrance for the processing of scientific spectra or the recorded format may even be refused by certain analysis software. However, for educational and advertising purposes such nice looking spectra can show at a glance the color gradient from red to blue and helps beginners to correctly align the recorded spectral stripe.

Ordinary digital compact and DSLR cameras, mainly designed for daylight photography, are mostly equipped with an IR filter, preventing a recording of the long wave spectral domain. However, such cameras may be quite suitable for the first attempts of beginners, for example combined with transmission or makeshift CD ROM gratings. For some DSLR cameras a debayering of the sensor is possible as well as the removal of the IR/UV blocking filter, also called a low-pass filter. The latter technique is called the full spectrum (FS) modification. The technique of debayering is a delicate process whereby the color filter array (CFA) is scratched off from the sensor with the aid of an appropriate solvent. It is recommended to let a professional debayer your camera. Nevertheless, and at your own risk, on the Internet numerous step-by-step DIY instructions can be found. Unfortunately, depending on how the Bayer matrix has been produced by the manufacturer, not all camera series can be debayered. Regarding the different architecture of a complementary metal–oxide–semiconductor (CMOS), mostly used in a DSLR camera, and a CCD, binning to higher sensitive pixel blocks is not possible. A bigger disadvantage of using a DSLR is the missing cooling possibility, although existing solutions are available.

6.3 The Recording System: Telescope, Spectrograph and Camera

6.3.1 Preliminary Remarks

Only the combination of a spectrograph with a suitable camera turns a "spectroscope" into a real recording unit. Also important are the optical specifications of the telescope. All elements must finally fit together or have to be adapted in order for a specific astronomical task to be performed. Next we cover some important aspects and requirements for a proper interplay of this system. Special thanks to Martin Huwiler for his decisive contributions to this section!

6.3.2 Limiting Magnitude of the System

The following factors affect the achievable limiting magnitude of a spectrograph:

- The aperture D and the focal length f of a telescope. These values must fit to the optical design of the spectrograph and the focal ratio $N = f/D$ should be equal or at least close to that of the collimator optics. If N is smaller, the entire flux cannot pass through the collimator and light gets lost; if larger the whole grating will not become lit up and resolution is reduced (see Equation {4.22}). With DADOS, Baches [58], Lhires III [59] and SQUES [60] this ratio is $f/10$, with LISA $f/6$ and ALPY $f/4$ [59].
- The operating principle, the resolution R, the slit width and the optical quality of the spectrograph.
- The operating principle, the quality, the specific pixel raster and the selected binning mode of the recording camera.
- The object class: Objects predominately generating emission lines are advantaged compared to pure continuum emitters.
- Location of the observation site: The recording of very faint objects requires a very dark sky without strong moonlight and no cirrus clouds. For apparently point shaped objects the quality of the seeing influences the spot size and therefore also the flux of light passing through the slit. For brighter objects, however, apart from strongly illuminated urban areas, this fortunately plays a rather minor role for spectroscopy.

6.3.3 Exposure Times for Grating Spectrographs

Among amateurs *SimSpec* Exposure Time Calculator (ETC) by Christian Buil is a well known, Excel based spreadsheet program, which, based on several parameters, allows the approximate determination of the required exposure times and the reachable limiting magnitude of a given spectrograph [75]. Independently to this what follows summarizes experiences made by the author of the *Spectral Atlas* [1].

For each setup (telescope/spectrograph/grating/camera) the required exposure times must be determined experimentally. Thereby, in the brightest area of the recorded picture, an intensity of approximately 2/3 of the reachable maximum saturation *must never* be exceeded; otherwise artifacts and even saturated emission lines may arise in the spectral profile. This intensity value is of course not achievable with *faint objects*. Analogously to astrophotography the noise can be reduced here by stacking of several pictures. Among others, the IRIS program is very well suited for this task.

A remark for perfectionists: For extremely faint objects, significant compromises in terms of signal-to-noise ratio (SNR) are unavoidable. A SNR of ~100 is generally considered to be a typical useful value. However, in special cases and depending on the specific task, even in the professional field profiles with SNR of just ~10 are analyzed! Despite using available software functions a reliable value for SNR is not easy to determine.

The following exposure times always refer to the configuration C8/DADOS with 200L Grating/Atik314L+ recording single exposures of which several should be recorded and stacked. For very bright, ordinary stars, with predominantly continuum radiation, just a few seconds are needed. However, at an apparent magnitude of $m_v \approx 6$ up to several minutes are required. At least with fainter objects, it makes sense to significantly increase the sensitivity of the camera by the binning mode (e.g. 2×2). For DADOS, in combination with the 25 μm slit, a loss of resolution can hardly be observed this way. For other configurations, this needs to be clarified in each case (see Section 6.3.6).

To estimate and optimize the exposure time also the following effects need to be considered:

- Apparently point-shaped, faint objects are easier to record than such like nebulae, appearing two-dimensional. For the latter, with a given f-number of for example $f/10$, using a larger telescope the exposure time cannot be reduced significantly.
- Objects, emitting most of their radiation by a few discrete emission lines, need less exposure time than pure

Table 6.1 Sampling S for different slit widths B, assuming pixel size $P = 6.25$ μm and binning mode 1×1.

	f_{Ob} [mm]	f_{Co} [mm]	β	B [μm]	FWHM [μm]	Sampling, S [number of pixels]	Remarks/ Recommendation
DADOS	96	80	1.2	25	30	4.8	Oversamp./ 2×2 bin.
	96	80	1.2	35	42	6.72	Oversamp./ 2×2 bin.
	96	80	1.2	50	60	9.6	Oversamp./ 3×3 bin.
SQUES	75	120	0.625	15	9.38	1.5	Undersamp./
	75	120	0.625	20	12.5	2.0	~ adequate
	75	120	0.625	30	18.75	3.0	Oversamp.
	75	120	0.625	40	25	4.0	Oversamp.
	75	120	0.625	50	31.25	5.0	Oversamp./ 2×2 bin.
	75	120	0.625	60	37.5	6.0	Oversamp./ 2×2 bin.

"continuum radiators," like ordinary stars. This includes, for example, apparent brighter, point-shaped planetary nebulae, Wolf Rayet stars, novae, supernovae and the AGN of emission line galaxies of the type starburst, Seyfert and quasar [1].

- Due to the specific sensitivity of modern CCD cameras, stars of the late spectral classes often require less exposure time than the early ones.
- The sampling should be optimized by adjusting the binning mode to the slit width (Table 6.1).
- Particularly for very faint, point like appearing objects, the setting of the slit width should be adapted to the seeing conditions or the diameter of the spot size. In Figure 6.3 this relation is displayed for different optical parameters.

With this setup the limiting magnitude for apparently point-shaped emission line objects, proved to be $m_v \approx 14$ and for continuum radiation sources $m_v \approx 13$. However, this requires excellent conditions and further a stacking of several shots, each recorded with at least 30 minutes exposure time in the 2×2 binning mode.

Experiments have shown that an exposure time of just a few seconds (!) is needed, to record the brightest O III line of the seemingly point-shaped planetary nebula NGC 6210. Accordingly just a few minutes are needed to record the brighter members of this class [1].

A special case is the recording of the emission nebula M42 (Orion). Its exceptional brightness allows single exposures of less than one minute. In contrast, however, the recording of faint emission nebulae like M27 and M57 is very time consuming. Extreme examples are the supernova remnant M1 or the Crescent Nebula NGC6888. For these, to obtain an acceptable profile, at least 30 minutes are required for a single exposure in the 2×2 binning mode. This effect was recognized by the astronomers of the early twentieth century, such as Vesto Slipher. At that time to record a spectrum of M1 with the 100 inch Hooker telescope and the "fastest" available films required an exposure time of several nights!

To compare the brightness of apparent 2D objects like nebulae and galaxies and to roughly estimate the chance to record it, the surface brightness (expressed in magnitudes per unit area) of an object must be known. It is expressed as surface brightness (SB) either in magnitudes per square seconds [mag arcsec^{-2}] or per square minute [mag arcmin^{-2}]. Due to the logarithmic scale the conversion between the two units is very simple [91]:

$$\text{mag arcsec}^{-2} = \text{mag arcmin}^{-2} + 8.89 \qquad \{6.1\}$$

As a rough reference: The weakest emission line object, recorded with the above mentioned setup, during an absolutely cloud- and haze-free night sky, was NGC 6888, the Crescent Nebula, with a surface brightness of SB = 24.64 mag arcsec^{-2} = 15.75 mag arcmin^{-2}. The quality of the night sky did not allow a visual observation of this nebula, so the positioning of the slit had to take place based on the known

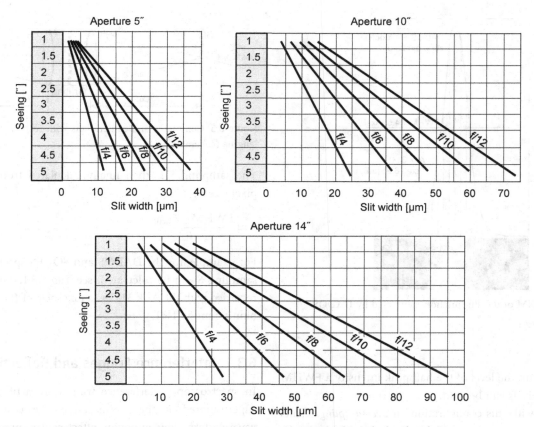

Figure 6.3 Fit of the slit width to the spot size of the telescope. Diagrams by M. Huwiler, calculated with Simspec

field star pattern. This spectrum and the associated recording details can be found in [1]. For comparison the SB values for M1 are 20.5 mag arcsec^{-2} corresponding to 11.6 mag arcmin^{-2}. It is important to note that between various data sources, differences up to half a magnitude may occur!

However, in most of the publications brightness values can be found where the average surface brightness per square unit has been converted to the size of a star. Such data are useless for this purpose. In addition, with reference to common stellar magnitudes, they suggest much too high luminosities for such faint seemingly 2D objects.

6.3.4 Pixel Size and Sampling with Slit Spectroscopy

Astrophotographers already know that with monochrome sensors, the smallest image detail – for example, the full width at half maximum (FWHM) of a stellar diffraction disk – needs to be covered and recorded by at least two adjacent pixels of the size P. Analogously to the signal theory the minimal sampling is called here the Nyquist criterion which is expressed as a number of pixels. For enhanced

requirements and SNR > 100, as a rule of thumb, the minimum sampling S must be increased here to ~2.5 P. This way the full information content of the spectral profile can be recorded. This value can empirically be reduced to ~1.6 P, if only the exact position of a line, for example for a radial velocity measurement, must be determined.

6.3.5 Determination of the Sampling by a Recorded Slit Image

In spectroscopy, the smallest detail to be recorded is also determined by the slit width B. For conventional slit spectrographs the sampling S, the number of pixels covering the slit width, can roughly be determined with a recording of the slit image. For this purpose the angle of the grating position must be adjusted until the slit image of the zeroth order is reflected directly onto the center of the image sensor.

If we enlarge a section of the image with the image editing program until the pixel structure becomes visible, the slit image no longer appears as a rectangle, but has a Gaussian-like brightness distribution – this is due to diffraction effects.

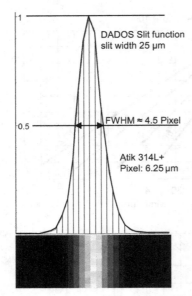

Figure 6.4 FWHM of the slit function, generated by DADOS 25 µm slit

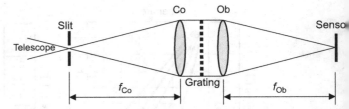

Figure 6.5 Collimator and camera lens with respective focal lengths (schematically)

The sampling S is then finally calculated from the known pixel size P:

$$S \approx \frac{\mathrm{FWHM}_{\text{Slit function}}}{P} \qquad \{6.4\}$$

For example, for the DADOS and SQUES spectrographs the sampling S can be calculated, see Table 6.1. The spectra are recorded with the Atik 314L+, a pixel size of $P = 6.25$ µm and binning mode 1×1.

6.3.7 Interference Fringes and Reflection Ghosts

In spectroscopy interference fringes are artifacts appearing superimposed on the continuum course as a fine, regular wave pattern. This annoying effect can occur in moderately high-resolution spectra with certain combinations of cameras and spectrographs and be verified if it becomes visible also on flat-field images. The causes are multiple reflections within the cover glasses of the CCD sensors, whose thickness determine the period of the "ripples."

This effect may also occur, with much coarser wave patterns, in low-resolution spectra. In this case the origin is suspected to be generated within the CCD sensor. For amaters, but also recommended by ESO-MIDAS, the most effective countermeasure has proven to divide the object image by a flat-field exposure taken with the very same instrument configuration.

Interference fringes must not be confused with "ghosts," normally appearing as bright spots in the recorded spectrum, which mainly originate from reflections between the CCD cover glass and the grating of the spectrograph. They are difficult to remove and reach a disturbing extent if the spots are nearly as, or even brighter than the recorded spectrum.

6.4 Recording of Echelle Spectra

6.4.1 Preliminary Remarks

The following explanations are based on and demonstrated by recordings of the SQUES echelle spectrograph [60], which is

In Figure 6.4, on the level of the half peak intensity a FWHM of about 4.5 pixels can be read.

Therefore, with this configuration an *oversampling* results and for average requirements the 2×2 binning mode can be applied here without any loss of resolution. Thereby the light falls on interconnected and therefore much more light-sensitive pixel blocks. The advantage here is that for faint objects the required exposure time can be reduced by more than half.

The graph of the slit image was generated here with IRIS and Vspec and the useless continuum intensity was subsequently subtracted. Finally the peak intensity has been normalized to unity as described in Section 8.2.6.

6.3.6 Analytical Determination of the Sampling

If the focal length of the collimator f_{Co} and camera lens f_{Ob} are equal, the original slit width B is displayed on the sensor approximately as the FWHM of the slit function in the ratio ~1:1. Figure 6.5 shows schematically collimator and camera lens with their respective focal lengths.

However, if $f_{\mathrm{Co}} \neq f_{\mathrm{Ob}}$, the scale factor β plays a role, and the displayed FWHM of the slit width B is given by the following approximate equations:

$$\beta = \frac{f_{\mathrm{Ob}}}{f_{\mathrm{Co}}} \qquad \{6.2\}$$

$$\mathrm{FWHM}_{\text{Slit function}} \approx B\beta \qquad \{6.3\}$$

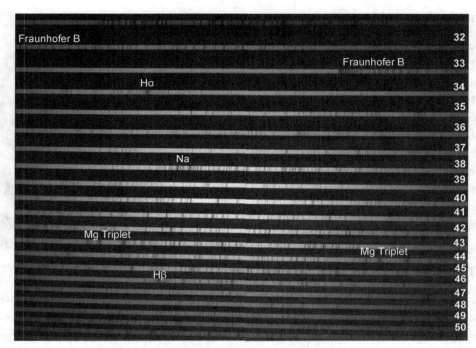

Figure 6.6 Orders 32–50 of the solar spectrum, recorded by SQUES echelle spectrograph

applied by one of the authors. Analogously adapted it also applies of course to other echelle spectrographs.

6.4.2 Special Features of Echelle Spectra

Compared with profiles generated with conventional grating spectrographs, the processing, evaluation and normalization of echelle spectra is much more demanding. The recorded image is called an echellogram. Separated by the cross disperser the high-resolution spectrum is here distributed to numerous stripes, which are called "orders," showing overlapping wavelength ranges. Figure 6.6 shows an excerpt from the solar spectrum. The orders 32–50 generated by the SQUES echelle spectrograph exhibit some striking spectral lines. So the overlapping is here clearly recognizable, for example on the top by the Fraunhofer B absorption or in the bottom third by the magnesium triplet (~λ5168–5185).

In contrast to the spectra recorded by BACHES or SQUES and depending on the optical design, the single orders generated by other spectrographs may run strongly curved, requiring specific software for processing and analysis.

6.4.3 The Orientation of the Spectral Image

The camera must be mounted on the spectrograph so that the following requirements are met:

- Due to the slight curvature of the stripe, the necessary horizontal leveling of the orders requires some care. If only one of them is to be analyzed, then this one is to be aligned as exactly as possible parallel to the horizontal edge. This adjustment can be made by the mounting angle of the camera. It is necessary to reach a reasonable compromise.
- The long wavelength, red region of the spectrum must be positioned on the top and the short wavelength, blue part on the bottom.
- Also for echelle spectra the convention requires for single orders the long-wavelength end to be to the right and the short-wavelength part to be to the left.

It should be noted that some software packages, such as Maxim DL and IRIS, may display a mirrored picture after loading! Further with the frequently applied Maxim DL for the storage of a picture the "compression type" must necessarily be set to "uncompressed," otherwise IRIS, for example, is unable to load the picture!

For beginners it is very difficult to orient themselves without any aids in highly resolved echelle spectra. Therefore in Appendix C a detailed, roughly wavelength calibrated solar spectrum can be found, which is designated for observation planning, showing for each order the covered wavelength range. Used before the observation session it enables one to predict the approximate location of a specific line in the spectrum – even for stars of other spectral types.

6.4.4 Focusing of the Spectral Image

In respect of the focus, the optical design (for example the BACHES and SQUES spectrographs) enables a good compromise, even in the peripheral zones, over the entire area within the orders 31–51. To record such a broadband spectrum, the best focus should be adjusted at approximately 1/3 of the order length and *not* in the center of the image.

If a short spectral range or just an individual line is to be investigated, *this detail* should be *focused* with extra care. This is best done previously with the help of the daylight spectrum. If possible the camera should run here in the loop mode with relatively short exposure times.

6.4.5 Exposure Times for Echelle Spectrographs

The exposure times required are significantly longer compared with a recording of the same spectral domain by a conventional grating spectrograph. The wavelength range, covered by the 200 L mm^{-1} grating of DADOS, is distributed, for example by the SQUES echelle spectrograph, to about 25 orders. This gives a first rough relative indication for the additionally required time effort! In practice, however, with the same primary optics and slit width, results in a factor of about 30. Anyway, the enlargement of the slit width and the increase in the binning mode of the camera, also reduce the exposure time significantly here. In the range of the shortwave orders the exposure time, compared with a recording in the red wavelength region, may be several times longer. For the recording of broadband spectra compromises are necessary.

The limiting magnitude of an echelle spectrograph depends particularly strongly on the considered wavelength domain. This effect is also caused by the dramatic drop in sensitivity of current astronomical cameras in the blue range of the spectrum. Based on the configuration C8/SQUES/Atik314L+ and on the slit widths and binning modes, listed in Table 6.1, the limiting magnitude for the domain of $\sim\lambda\lambda4500$–8000 is $m_v \approx 9$. Nevertheless, to record the Fraunhofer H + K Lines Ca II at $\lambda\lambda3934$ and 3968, it drops significantly to $m_v \approx 5$.

6.5 Influences of Mount and Guiding

6.5.1 Mechanical and Structural Problems with Small Mounts

From the mechanical and structural point of view, spectroscopy is nothing else than a special form of astrophotography with a significantly increased weight. The only difference

Figure 6.7 The spectrograph, an additional load to the mount

being the addition of a spectrograph which is connected between telescope and camera (see Figure 6.7). However, its weight and dimensions aggravate the structural problems of small mounts, which requires adequate solutions. Furthermore, a flip mirror is usually screwed in front of the spectrograph and the slit is observed with an additional camera. This is operationally necessary, but the cantilever moment is increased because the lever arm is extended by the size of the additional flip mirror. Further it is increased also by the weight of the flip mirror and the additional camera for the observation of the slit. The whole assembly is mounted on the telescope as a console. Anyway the resulting bending moment, acting on the 2″ adapter, causes no further problems for most of the larger telescopes ($\geq 8''$).

This applies also to the screwed T2-connections. Plugged connections, however, tend to be a weak point and must be constructed to resist bending and with little play. The appropriate clamping screws must be tightened well to avoid shifts within the adapter, caused by telescope movements. The same applies, if adjustable, to the locking screws of the grating angle (e.g. DADOS).

Since the whole setup structurally acts as a console, even a slight bending of the spectrograph casing is inevitable. The extent depends on the bending stiffness of the whole structure and the elevation angle of the telescope. This, for example, may have consequences on the accuracy of a calibration with a light source (see Section 8.1.3).

6.5.2 The Option of Fiber Coupling

All these problems can be avoided if the spectrograph is coupled to the telescope just by an optical fiber. Separated this way the device may be mounted in an air-conditioned or even a vacuum box. This setup is mainly applied in the professional field to enable undisturbed high precision measurements of extremely low radial velocities, as typically occur in the search for exoplanets. Further, for point sources by fiber coupling a small shifting of the spectrum on the detector, by an unsymmetrical illumination of the slit, can be avoided or at least diminished [4]. Typical drawbacks of such a coupling are a loss of resolution and light, reducing the limiting magnitude of the spectrograph and also a non-adjustable, fixed slit width, determined by the diameter of the fiber. A typical representative in the amateur field is the echelle spectrograph "eShel" by Shelyak Instruments [59].

Figure 6.8 Adjustment of the load distribution: C8 on a *Vixen Sphinx Deluxe* mount

6.5.3 Impact of the Spectrograph Load to Small Mounts

Depending on the orientation of the telescope, the additional load of the spectrograph mainly produces different strengths of torque, acting on the declination and right ascension drive of equatorial mounts. This additional moment may even become zero for certain orientations. The alt-azimuth mounts behave somewhat differently here. In any case, suitable counter-measures are required.

6.5.3.1 Equatorial Mounts

The simplest cases are equatorial mounts with a dovetail rail. If extended beyond the end of the telescope for almost all orientations an appropriate load distribution can be adjusted. However, this requires careful planning because this load adjustment must be done at the very beginning of the observation session, before the alignment of the telescope takes place. It has also proven succesful, for a certain setup and elevation angle, to carve position marks on the rail in order to quickly reproduce a balance point. To adjust the load distribution, the telescope will be best aligned horizontally to minimize the risk of sliding out (Figure 6.8).

6.5.3.2 Alt-Azimuth Mounts

Generally an alt-azimuth mount is suitable for spectroscopy because the field rotation, at least for objects with a puncti-form appearance, plays no role. Because the fastening of the telescope is here mostly not adjustable the adaptation of the load balance is significantly more difficult. Either awkward balance weights and/or special solutions are required, which are associated with a mechanical modification of the mount. To reduce here at least the lever arm, the spectrograph may be mounted perpendicular to the optical axis, by a rigid 2″ diagonal mirror.

However, best suited for spectroscopy are here the bulkier, mostly professional telescopes with Nasmith focus, allowing even heavy spectrographs to be mounted. Here, by a third (tertiary) mirror the beam is deflected horizontally through the supporting bearings of the altitude axis.

6.5.4 The Mounting Angle of Spectrograph and Cameras

Normally the spectrograph, when fastened to telescopes with small equatorial mounts, should be rotated and fixed in the socket so that the slit gets aligned with the declination slew. This way the image of the object can be moved with the hand control either parallel or perpendicular to the slit axis.

An exception being for long slit measurements of the rotation velocity the slit axis must be aligned parallel to the equator of the planet or to the edge of edge-on galaxies. In the latter case the known position angle (PA) of the longitudinal axis, related to the equatorial coordinate system, allows the spectrograph to be installed with an approximately correct mounting angle during the start up of the telescope (Figure 6.9). For galaxies the PA can be found in the NASA Extragalactic Database (NED; [26]) and is measured, analogously to the binary stars, counterclockwise, starting from the north direction. Thus a PA of 90° means that the edge runs parallel to the celestial equator (e.g. M104, Sombrero: PA = 89°).

The slit camera must be rotated and fixed in the socket so that the slit is displayed sharply and its axis is aligned parallel to the vertical edge of the screen.

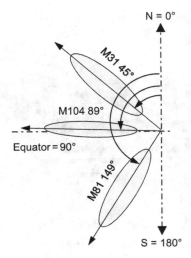

Figure 6.9 Position angle of galaxies

The recording camera must be mounted at the spectrograph so, that the following conditions are met:

- The spectral stripe to be analyzed must run as accurately as possible parallel to the horizontal edge of the image.
- According to convention the long wavelength, red domain of the spectrum must always be positioned on the right and the short wavelength or blue part on the left.

The image needs to be focused as accurately as possible with the camera lens. This can, for example, be achieved by means of the finely structured daylight spectrum.

6.5.5 Documentation of the Setup

In most cases, amateur astronomers use their cameras, flip mirrors and other accessories for both astrophotography and spectroscopy. In such cases, it is highly recommended to document each setup in detail. Thus, after long breaks in observation activities, or when switching during a session from one application to the other, unnecessary loss of time and trouble can be avoided.

6.5.6 Positioning of Faint Objects on the Slit

The slit mirror of a spectrograph shows just a very small section of the sky, which hampers the locating and positioning of the object, but reduces also the number of possible field stars for autoguiding. With the C8 (focal length of approximately 2000 mm) and the DADOS spectrograph, this is just ~8×8′. With larger telescopes and longer focal lengths, this area gets correspondingly smaller. The application of focal reducers is only useful if the collimator of the

spectrograph has been dimensioned for a corresponding focal ratio (see Section 6.3.2).

The following measures facilitate the proper positioning. In most cases a flip mirror in front of the spectrograph is indispensable. Combined with appropriate eyepieces, it shows a much larger section of the sky than the slit mirror. It allows a first rough positioning of the object in the center of the field of view (FOV). It makes sense to determine the diameter of the FOV for the configuration used. This value is also needed to prepare a finder chart. It can be estimated by comparing the visible section of the sky with a star map or measuring the transit time t [s] of a star with known declination δ by switched off guiding function:

$$\text{Field of view [arc seconds]} \approx t\,15\cos\delta \qquad \{6.5\}$$

For faint objects it is strongly recommended to prepare a finder chart. So, the field star pattern allows an easy localization of the object in the FOV. The chart should be correctly parameterized in order to display the appropriate section of the sky, the suited limiting magnitude of the stars, as well as considering possible mirroring effects of the optics. To prepare the finder charts, the corresponding application of the AAVSO website is highly recommended [24]. If in the vicinity of the target object only a few and/or faint field stars are present, it is recommended to prepare two charts – one with the exact and another one with an approximately double sized FOV of the flip mirror optics.

For a precise positioning of the object on the slit, the mount must provide a correspondingly fine control. As a rough reference, the *Vixen Sphinx Deluxe* and some 2000 mm focal length, this task requires the highest zoom level provided by the Starbook software.

6.5.7 Specific Requirements to the Guiding Quality

Like astrophotography with long exposure times spectroscopy also requires a correction of the automatic tracking. However, this is normally not performed by parallel telescopes or off axis guiders but by the slit observation optics either visually, combined with manual corrections, or electronically with a guiding camera and software supported autoguiding. Here, only the spectroscopic peculiarities of this topic will be considered. Excluding the high precision measurement of radial velocities, for the recording with slit spectrographs of objects appearing as punctiform, the requirement for the tracking accuracy is lower as for astrophotography, because the optics of the telescope do not need to directly produce a sharp image. While recording, only this light of an object is integrated by the image sensor, which passes through the slit of the spectrograph. The slit width is one of the determining parameters for the resolution power. Therefore, the slit spectroscopy is pleasingly insensitive to the seeing quality, which for most amateur applications, mainly affects the exposure time required to achieve a particular SNR. Unfortunately this is not the case for recordings with slitless applied transmission gratings.

6.5.8 Load Distribution and Autoguiding Process with Small Mounts

Astrophotographers have long recognized that with small mounts autoguiding works poorly, or even not at all, if the balance of the telescope is set too accurately. Mostly concerned here is the declination drive of small equatorial mounts, with a rather inaccurate polar alignment. The challenge is to set a reasonable strength for this imbalance, which is only valid for a certain section of the sky. It is important to note that even at long exposure times these conditions must be fulfilled within the entire tracking range. Critical are certain orientations close to the zenith where, with smaller mounts, the desired imbalance cannot be achieved just by mere shifting of the load and/or counterweights.

6.5.9 Autoguiding: Interaction of Hardware and Software Components

In principle the autoguiding process is quite easy to understand. However, in practice it very often turns out as quite complex, because several software and hardware components have to interact properly. Therefore, if an acquisition of new equipment is planned, it is highly recommended to choose an already proven combination of telescope mount, software and guiding camera. The latter has a double function, because in spectroscopy it is most often additionally used for the observation of the star on the slit. It is further indispensable for the recording of faint objects because, by extending the exposure time, weak field stars, galaxies and nebulae also appear on the slit mirror.

There is a choice between several existing guiding software packages (mostly freeware). Anyway, it is essential to clarify whether the operation of a particular camera is supported. Most guiding applications communicate with the so-called ASCOM platform. This software must be installed, together with the mount-specific driver plugins, on the same computer, which in most cases can also be used for recording of the spectra.

Another, sometimes problematic, interface exists between the computer with the guiding software and the electronics of the mount. In addition, today autonomous solutions are also available, operating independently from external computers, but in certain cases may also cause interface problems. Another hurdle is the parameterization of the guiding software. It is impossible to propose here any recipes because it depends on too many factors. It just remains to consult the manuals, in combination with your own experiments.

6.5.10 Spectroscopic Aspects of the Guide Star

The best case is a separate nearby field star on the slit mirror, which can be used as guide star. In frequent cases if a very faint target object must at the same time be applied as a "guide star," it may be fully covered by the slit. In this case

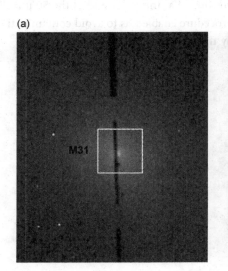

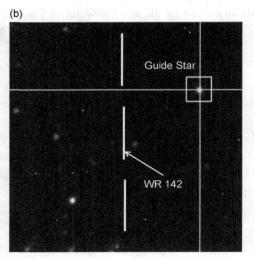

Figure 6.10 PHD Screenshots, (a) M31 and (b) WR 142

the autoguiding software PHD guiding has proven to be pleasingly insensitive. It triggers of course a "guide star lost" alarm, but as soon as the object reappears the crosshairs catch it again.

Figure 6.10(a) shows a PHD guiding screenshot of the core of the Andromeda galaxy M31 on the 25 μm slit of DADOS. Used as a "guide star," this setup caused some troubles here. Figure 6.10(b) shows the very faint WR142 (13.8^m) fully hidden behind the slit and further the field star, applied for autoguiding (slit camera: Meade DSI II color). This configuration caused no problems and allowed an acceptable tracking accuracy. The applied exposure time for both cases was 3 s.

6.5.11 Recording of Close Binary Star Components

The recording of close binary star components offers special challenges. The critical distance depends on various parameters and conditions, such as focal length, quality of tracking, seeing etc. Here follow some experiences with a C8 and the Vixen SXD mount.

Binary stars with large angular distances, such as Albireo, cause no specific problems (Figure 6.11(a)). Here, the second component, in a comfortable distance of 34″, may even be used as an appropriate guide star.

In the binary star system α Cnv (Cor Caroli) the angular distance of 19″ between the components is significantly smaller (Figure 6.11(b)). The brighter A-component ($m_v = 2.8$) is here on the slit and acts additionally as a guide star (PHD autoguiding). Under good conditions this is not really a problem. However, in bad seeing conditions or combined with thin cirrus clouds, after a short-term disturbance of the visibility, the autoguiding may jump on the wrong star. A possible countermeasure may be the size-reduction of the guide frame.

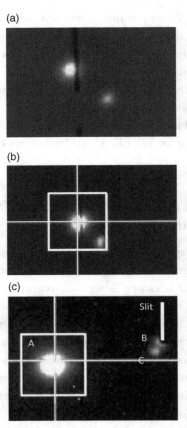

Figure 6.11 (a) Albireo, (b) Cor Caroli, (c) 40 Eridani B, screenshots PHD Guiding

Rather critical is the recording of the white dwarf 40 Eridani B ($m_v = 9.5$). From the somewhat fainter C-component ($m_v = 11.2$) it is separated just by 9″ (Figure 6.11(c)). The A-component ($m_v = 4.4$) is here the guide star at a comfortable distance of 83″. To record the B-component, it was positioned at the lower end of the 50 μm slit (DADOS). This procedure enabled us to avoid contamination of the spectrum by the very close C-component.

7   Processing of Recorded Spectra

7.1 Available Software for Data Reduction

The next step after a successful recording of the spectrum is data reduction. Several software packages are available. The most important task here is to convert as perfectly as possible all spectral information into a graphical presentation, which can then be used to analyze astrophysical properties and behavior of the astronomical object of interest. As outlined in this book, many astrophysical stellar properties can be derived from the spectra, for example, the effective temperature, pressure, gravity and element abundance. Other

information concerns expanding material, ejected by stars due to different mechanisms, electromagnetic fields or rotation velocity, always keeping in mind that spectroscopy is giving information gathered in the line of sight.

7.1.1 Software Packages for the Amateur

Table 7.1 presents an overview of the software packages actually available for amateurs. The entry level program RSpec [72] is the only fee-based software, but is fortunately reasonably priced. All other software packages are freely

Table 7.1 Software packages for amateurs

Software	Author	OS	Type	Info
BASS project	John Paraskeva	Windows	Reduction Analysis	**Basic Astronomical Spectroscopy Software** https://uk.groups.yahoo.com/neo/groups/ astrobodger/info
IRIS	Christian Buil	Windows	Image Processing	Part of Audine project by Aude (Association of the **U**sers of Electronic **D**etectors) http://www.astrosurf.com/buil/iris/iris.htm
ISIS	Christian Buil	Windows	Reduction Analysis	**I**ntegrated **S**pectrographic **I**nnovative **S**oftware http://www.astrosurf.com/buil/isis/isis.htm
RSpec	Tom Field	Windows	Reduction Analysis	**R**eal-time **Spec**troscopy http://www.rspec-astro.com/
Spc Audace	GNU-Cooperation	Windows, Unix, Linux, Mac OSX	Reduction Analysis	Part of Audela project, open source astro-imaging software http://bmauclaire.free.fr/spcaudace/
Spectroscopy Analysis Software	Ken Whight	Windows	Analysis	http://www.thewhightstuff.co.uk/?page_id=128
VSpec	Valerie Desnoux	Windows	Reduction Analysis	**V**isual **Spec** http://www.astrosurf.com/vdesnoux/

available. The BASS project [76] needs an additional free online registration via the Astrobodger website. All websites offer free downloadable manuals and/or instructive videos and illustrated tutorials. In this book we demonstrate the image processing, reduction and analysis by the use of the programs IRIS [71] and VSpec [74]. Similar results can be obtained with the other programs.

7.1.2 Professional Software Packages

Table 7.2 gives an overview of software packages freely available for professional astronomers. These software packages are used by professional observatories, institutes and academics. These programs are referred to as standard reduction and analyzing tools in scientific published papers. When used by the advanced amateur an intensive course on their use is highly recommended. Moreover, ESO-MIDAS and IRAF operate through mainly commando-based instructions, which require a good knowledge of the program commands. Nevertheless they are the most powerful astronomical software available and capable of avoiding artifacts, known as the feared "black box" phenomenon.

7.1.3 Spectroscopy @ Cyberspace: The Virtual Observatory

Stellar spectral databases have been growing through the years. Based on observed high resolution stellar spectra and developed theoretical astrophysical atmospheric models, synthetic stellar databases were generated. To improve the interchangeability of the stellar spectral files worldwide a universal standard protocol was proposed: the simple spectra access protocol (SSAP). To consult, analyze and interpret those database files a special tool defined as the virtual observatory (VO) was developed. The International Virtual Observatory Alliance (IVOA) was founded in 2002. Cooperation between currently 19 countries resulted in a user friendly, virtual observatory interface. An example is the Euro VO AIDA/WP5 project by the European Virtual Observatory (Euro-VO), which formed the basis for the development of the well-known program Stellarium and the digital atlas Aladin. The use of a virtual observatory interface for the amateur is a powerful tool to gain a more profound understanding of the astrophysical properties of their own recorded spectra. Synthetic spectra can be transformed in simulated observations. The use of programs as SPLAT-VO, VOSpec and VO-SPECFLOW are scientific tools for the more advanced amateur (see Table 7.3; [77]).

The use of the virtual observatory (VO) gives access to profile diagrams based on atmospheric models. Measurements of equivalent widths (EW) or shifts of specific spectral lines used as indicators of astrophysical parameters combined with synthetic profiles are modern and powerful methods for determination of the stellar effective temperature T_{eff}, surface gravity (log g), abundances, micro- and macro-turbulence velocities (V_{mic}, V_{mac}) and rotation (V_{sini}). Combined with empiric methods this advanced spectroscopic data analysis is

Table 7.2 Professional software packages

Software	Author	OS	Info
ESO-MIDAS	ESO	Unix/Linux Windows (Virtual) Mac OS X	http://www.eso.org/sci/software/esomidas/ European Southern Observatory-Munich Image Data Analysis System
IRAF	NOAO	Unix/Linux Mac OS X	Image Reduction Analysis Facility http://iraf.noao.edu/
SPLAT	Peter Draper, Mark Taylor, Margarida Castro Neves	Unix/Linux Windows Mac OS X	Spectral Analysis Tool http://star-www.dur.ac.uk/~pdraper/splat/ splat-vo/splat-vo.html

Table 7.3 Virtual observatory software packages

Software	Author	OS	Info
SPLAT-VO	GAVO	Unix/Linux Windows Mac OS X Online	Spectral Analysis Tool – Virtual Observatory http://star-www.dur.ac.uk/~pdraper/splat/splat-vo/
VOSpec	ESA	Unix/Linux Online	http://www.sciops.esa.int/index.php?project=SAT&page=vospec
VO-SPECFLOW	LUPM	Online	http://bass2000.bagn.obs-mip.fr/vospecflow/index.php

commonly used in professional astronomical spectroscopy, but can also be used for educational purposes by advanced amateurs.

7.2 From the Recorded Spectrum to the Calibrated Intensity Profile

7.2.1 Preliminary Remarks

In contrast to astrophotography any "cosmetic" treatment of recorded spectra by Photoshop, for example contrast enhancements and the like, is strictly forbidden and would inevitably destroy the scientific content. Exceptions are the removal of hot pixels and some further necessary process steps as described later. Further, as previously sometimes recommended broadening of the spectral stripe while recording, for example by movements with the declination drive and the like, is no longer necessary today.

This procedure is roughly displayed in Table 7.4, as processed by the IRIS and VSpec software. Analogously adapted it applies of course to other programs. The processed pictures show the white dwarf 40 Eridani B, recorded by C8, DADOS 200 L mm^{-1} and ATIK 314L+ [1].

7.3 Removal of Light Pollution and Airglow

7.3.1 Objects of Point-shaped Appearance

The recording of an *object* of *point-shaped appearance* generates a slim "spectral-stripe," whose height depends mainly on the seeing and autoguiding quality. The light pollution can here simultaneously be subtracted with the sky background (e.g. by IRIS). The following example shows the recorded spectrum of WR142 (hot pixels already removed). In the upper stripe of Figure 7.1 the light pollution is easily recognized by the emission lines passing through the entire height of the slit. The four crosses (IRIS) indicate the selected

Recorded Raw Image: Night Sky- plus WR142 Spectrum

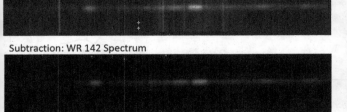

Subtraction: WR 142 Spectrum

Figure 7.1 Subtraction of light pollution from a point-like object (WR 142)

reference ranges to be subtracted from the finally desired spectrum of WR142. In the lower stripe, as a result of the subtraction, the remaining spectrum from WR142 is displayed.

7.3.2 Objects Appearing as 2D

The following example (Figure 7.2) shows the supernova remnant M1, recorded with the C8, the camera Atik 314L+ in 2×2 binning and an exposure time of 30 minutes. This object appears here so large that it spreads over the entire slit height (upper stripe). Therefore, in this picture no reference can be obtained to subtract the sky background and light pollution, appearing here in comparable intensity to the desired M1 spectrum. Therefore, the light pollution and airglow was recorded separately, just somewhat outside the nebula (middle stripe). Finally, as a result of the subtraction (here with Fitswork), in the bottom stripe just the remaining spectrum of M1 is seen. Striking here is the appearance of the characteristic split and deformation of all emission lines. It is important to timely record the sky background as close as possible and under the same conditions as the spectrum of the object.

7.4 Removal of Remaining Hot Pixels and Cosmics

Randomly recorded cosmics or remaining hot pixels in the area of the recorded spectral stripe must be removed. Otherwise false emission lines may be simulated this way. If such artifacts appear within the stripe the removal requires special care. So, exclusively the concerned pixels should be canceled. Otherwise in certain cases false absorptions may be recorded! Very convenient for this task is the according function of the IRIS software.

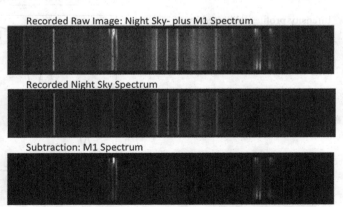

Figure 7.2 Subtraction of light pollution from an object of 2D appearance (M1)

Table 7.4 Steps from the recorded profile to the calibrated intensity profile

Process	Display
Stacking of two recorded object pictures with IRIS: Exposure time: 2×30 min, 2×2 binning mode. Sky and dark are subtracted, remaining hot pixels are removed.	
Calibration spectrum, generated by the modified Glow Starter Relco SC480, recorded with the same setup 1×20 sec, 2×2 binning mode.	
1D spectrum of the object, generated by IRIS by averaging of the brightness values within the individual vertical pixel columns.	
1D calibration spectrum generated by IRIS by averaging of the brightness values within the individual vertical pixel columns.	
2D intensity raw profile, generated by Visual Spec (VSpec), based on the gray values of the 1D spectrum.	
Superposition of the raw profile with the calibration spectrum (here Relco SC480, dotted line). Selection and designation of some suited calibration lines to be processed.	
Intensity profile, calibrated by wavelength.	

Table 7.4 (*cont.*)

Process	Display
This last step depends strongly on the purpose of the spectrum. Here the most frequently used rectification of the profile is shown, achieved by division of the pseudo-continuum by its own course. For details and further options refer to Section 8.2.	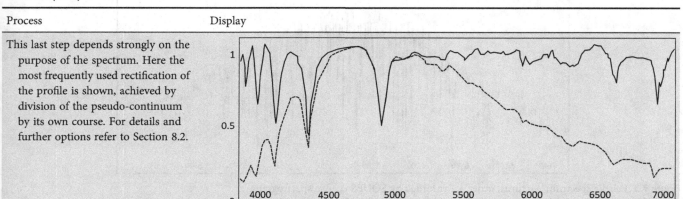

7.5 Dark-frames and Flat-fielding

These processes are extensively described in numerous publications about astrophotography. Here follow just some remarks to spectroscopically relevant peculiarities. In amateur circles this topic is discussed frequently and is very controversial.

7.5.1 Dark-frames

For longer exposure times the subtraction of dark frames for the dark current correction mostly makes sense. Further, a big part of the hot pixels are removed this way. The randomly occurring cosmics, however, must be separately removed.

The application of the frequently presented standard reduction procedures, also called "pipelines," usually protects against criticism but do not necessarily show the best results. For amateurs, limited to some special cases, it may be worthwhile to experiment – so, for example, the data reduction of the recorded supernova remnant M1 (Figure 7.2) with an exposure time of 30 minutes per single shot (CCD temperature –20 °C). Here, the direct subtraction of the light pollution on the level of unprocessed raw images showed finally the best results. After the subtraction, the numerous remaining hot pixels together with the randomly recorded cosmics, have been removed "manually" with IRIS in order to prevent the generation of false "emission lines."

7.5.2 Flat-fields

Compared with astrophotography the flat-field technique has the following consequences on the recorded spectrum.

In the fields of astrophotography and photometry it compensates for the different sensitivities of the individual pixels and further the typical drop in brightness at the edges of the image (vignetting). In spectroscopy, the course of the recorded pseudo-continuum becomes additionally deformed, but never really "corrected," by the specific radiation characteristic of the flat-field light source. The calibration or normalization of the intensity is achieved here by other methods (see Section 8.2). With modern astronomical cameras the different sensitivity of the individual pixels plays a minor role here. Furthermore, within a user-defined height of the recorded stripe the intensity of each and every vertical pixel row gets "smoothed" by the specific algorithms of the data reduction (see Section 7.2).

In astrophotography the flat-field technique eliminates artifacts due to dirt and dust on the surface of the CCD sensor. However, for modern astronomical cameras this is barely a topic anymore and excluding cases of extreme accumulation of dirt, this has hardly an impact on the quality and course of the spectral profile.

However, important for spectroscopy, the flat-field technique acts against the phenomenon of interference fringes (see Section 6.3.7).

In professional astrospectroscopy, flat-fields are generally processed. For amateur purposes, this is only indispensible if the spectral profile is affected by artifacts of interference fringes. Further it is requested by certain software packages for the broadband processing of entire echelle spectra (see Section 7.6.1).

7.6 Processing of Echelle Spectra

7.6.1 The Processing of an Entire Echelle Spectrum

The overall processing of an entire echelle spectrum is highly demanding and requires the proper assembly of numerous

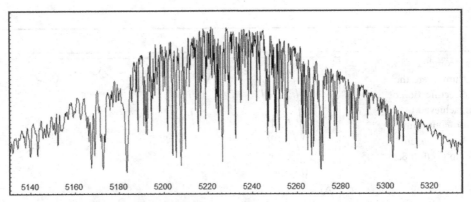

Figure 7.3 Echelle spectrum Arcturus, order 43, recorded by SQUES echelle spectrograph

orders. This can just be done with appropriate software support, such as MIDAS [69], IRAF [70] or ISIS [71]. It is mandatory mainly for echellograms, recorded by spectrographs generating more or less strongly curved orders. This procedure is described in the manuals and also required if the recorded spectrum has to be uploaded to scientific databases.

7.6.2 The Processing of Individual Orders

In most cases, echelle spectra are obtained for the high-resolution investigation of specific spectral lines. Here, at least for amateur purposes, the much simpler processing of individual orders is mostly sufficient. This requires spectrographs, generating straight or at most slightly curved orders with spectral lines which are aligned reasonably perpendicular. Therefore, individual orders, generated by BACHES or SQUES spectrographs, can conventionally be processed like spectra of ordinary grating spectrographs, for example with IRIS and VSpec (Section 7.1). A condition, however, is that the specific task requires just a rectified spectrum (Section 8.2.6).

Due to the specific optical design a common characteristic of all echelle spectra is a clearly visible fading of intensity towards the peripheral zones. Therefore all orders show bell-shaped pseudo-continua which are in no way connected with any radiation pattern of a recorded object. Adequate for most applications, at least for amateur purposes, is a rectification of the continuum intensity (Section 8.2.6). Figure 7.3 shows as an example of order 43 in the echelle spectrum of Arcturus.

If a spectral feature under investigation lies in the peripheral part of an order, two subsequent orders must be processed and subsequently be assembled. This is also enabled by the VSpec software. For detailed descriptions refer to the appropriate software manuals.

7.6.3 Subtraction of the Sky Background and Light Pollution

In echelle spectra due to the very short slit length, a direct subtraction of the sky background with the light pollution is impossible. If necessary, this can only be obtained by subtracting of a separately recorded, pure sky spectrum. For the removal of telluric H_2O and O_2 absorptions refer to [1] and the manuals of the data reduction software.

CHAPTER

8

Calibration of the Spectra

8.1 Calibration of the Wavelength

8.1.1 Preliminary Remarks

Usually spectra are plotted as the spectral flux density or intensity I over the wavelength λ. Absolutely necessary for most applications is the calibration of the wavelength axis. The recording of the spectra with electronic image sensors significantly facilitates this process. It enables us to assign the wavelengths to each and every vertical pixel column along the linearly scaled λ-axis and to determine the corresponding dispersion in Å/pixel.

There exist two basic options – relative calibration, based on standard air wavelengths of known lines and absolute calibration with light sources. Essential for an analysis is to know which method has been used to calibrate the profile (see Section 8.2.10 and Table 8.1)

8.1.2 Relative Calibration Based on Known Lines

This procedure is very easy and moreover recommended for the first calibration attempts by very beginners. Anyway this method is indispensable, if in a profile unknown lines should be identified, based on their rest wavelengths (see Section 9.3). For stellar spectra, obtained by transmission gratings without a slit, it is even the only option, apart from a very rough calibration considering the position of the zeroth-order relative to the spectral lines, and the known dispersion of the device.

8.1.3 Absolute Calibration with Light Sources

This principle is very simple. As a reference for the wavelength λ, the spectrum is superimposed with the exactly known emission lines of a light source, in most cases recorded separately and subsequently merged with the object spectrum by the analysis software (Section 7.2). This method is mainly applied to high-resolution spectra. It is mandatory for absolute, wavelength dependent measurements, for example the Doppler shift, or for stellar profiles of late spectral classes with only diffuse and/or with mainly unknown lines. This procedure is strictly limited to devices with a slit in front of the dispersing element.

8.1.4 Absolute Calibration by Unshifted Wavelengths of Atmospheric H_2O Lines

The unblended or the unshifted, telluric water vapor lines are often used for the calibration of highly resolved spectra, particularly around the Hα line. Details can be found in [1].

8.1.5 Linear and Nonlinear Calibration

The wavelength scale of a raw spectrum, obtained by a prism, runs in any case significantly nonlinear. Recorded with a modern transmission or reflection grating the linearity of the wavelength scale is much better but still far from perfect. So a linear calibration, performed by just two calibration lines, is clearly restricted to very short profile sections and very low requirements of accuracy. If at least three calibration lines show up within the relevant section, in any case a nonlinear calibration should be performed. Today these procedures are supported by analysis software (Section 7.1) and are well documented in the according manuals.

Table 8.1 Tasks and required calibration procedures (R = required, B = best option, P = possible option)

Methods of Calibration and Normalization Task	λ-calibration by rest wavelengths of known lines	Absolute λ-calibration by calibration light source	Pseudo-continuum of the raw profile	Rectified continuum, normalized to $I_C = 1$	Relative flux calibration by synthetic model star	Relative flux calibration by recorded standard star	Emissions scaled to theoretical Balmer decrement	Absolute flux calibration
Relative measurement of a wavelength difference Δλ	R	P	P	B	P	P		P
Absolute measurement of the Doppler shift	P	R	P	B	P	P		P
Measurement of FWHM	P	P	P	B	P	P		P
Measurement of EW	P	P	P	R	P	P		P
Determination of the spectral class	B	P	P	B	P	B		B
Line identification	R	P	P	B	P	P		P
Optical intensity comparison of the absorption lines	P	P		R				
Measurement of the reddened Balmer decrement	B	P		R				
Estimation of the original I_E- and I_A-values in the unreddened original profile	P	P			B	P		P
Comparison of the absorbed fluxes at different absorption lines	P	P			B	P		P
Evaluation of the emission intensities (e.g. for plasma diagnostics)	P	R					R	
Estimation of the effective temperature T_{eff} based on continuum slope	P	P			R	R		
Fluctuations of the continuum radiation	P	P				R		
Processing of profiles for certain scientific databases	R	R			R	R		
Processing of profiles with diffuse continuum-course, e.g. novae and SN.	R	R			R	R		
Flux measurement in physical units	P	P				R		R

8.1.6 Practical Aspects to Minimize Sources of Errors

The calibration procedure must be performed so that the influence of possible sources of errors will be minimized as far as possible. For an accurate calibration the following items must be observed.

- Between the recordings of the investigated object and the calibration spectrum, the setup of the equipment including the telescope must never be changed otherwise a detectable calibration error may occur. A change of the target bearing of the telescope also influences the acting bending forces on the casing of the spectrograph, which finally influences the recorded profile.
- Temperature variations cause a deformation of the grating size, as well as of the housing. Further, an additional shift is generated by a minimal change in the speed of light, shifting the wavelengths of the calibration spectrum relative to the object spectrum. Changes in air pressure have a similar influence. Therefore the calibration spectrum must be taken as close as possible to the object spectrum. For precision measurements even two calibration spectra are recorded and subsequently averaged – each recorded before and after the recording of the object spectrum. Applying the program *SpectroTools* by P. Schlatter [73], the dependence on temperature and air pressure of a λ measurement can impressively be demonstrated.
- From the same reasons before the measurement, enough time must be allowed for a proper adaptation to the ambient temperature for the entire setup: *telescope – spectrograph – camera*.
- The camera should be mounted so that the investigated spectral stripe already runs as horizontally as possible and thus an alignment, for example with the IRIS software, is no longer necessary. This way possible artifacts generated by rotation of the recorded spectrum can be avoided.
- The spectral line under investigation should not be located at the edge of the stripe but rather within a reasonable number of well-distributed calibration lines!
- All process steps, applied to the object spectrum, must analogously be applied to the calibration spectrum. Particularly the vertical extent and exact position of the binning zones must be identical for both stripes (e.g. IRIS).

8.1.7 Heliocentric and Geocentric Corrections

Relative to the Sun our observation site performs a very complex movement. It is determined by the Earth's rotation and the orbit of the Earth–Moon system around the Sun, which has a mean velocity of ~30 km s^{-1}. So it seems obvious that this phenomenon may have a considerable impact to measured radial velocities or may cause an unwanted red– or blueshift in the recorded spectra. A proven way out of this problem is a so called heliocentric correction, virtually transforming the obtained measurements as taken at the center of the Sun. Sometimes an appropriate time correction of an observed cosmic event is also necessary. Despite the fact that our solar system orbits the galactic center within ~250 million years, for most applications the center of the Sun can be assumed as a kind of virtual "fixed point." Prior to heliocentric corrections it is indispensable that the influence of the distance from the location of observation to the Earth's center is taken into account. This is known as the geocentric correction. By using telluric lines as a calibration tool a good approximation of the geocentric correction is reached. Fortunately today these highly complex calculations are supported by most of the available analysis software (e.g. IRIS, ISIS, Visual Spec or BASS). Mostly required are the geographic coordinates of the observation site, the equatorial coordinates of the observed object, as well as date and exact time of observation.

8.1.8 The Selection of the Calibration Light Source

The main criterion is to achieve a number of well distributed lines within the examined spectral range. The formerly popular neon glow lamp produces a narrow line grid, but unfortunately mainly in the red region of the spectrum. This solution is now obsolete, because we have developed low cost alternatives with glow starters, modified as gas discharge lamps [1]. So the model Relco SC480 produces approximately 270 evaluable lines (Ar, Ne, He) in the whole optical spectral range – for details see [1].

Highly resolved echelle spectra require calibration light sources generating a dense grid of emission lines in the entire optical range. For such highly resolved spectra in the professional sector, costly hollow cathode lamps, for example with iron and argon, powered with high voltage, are used. The modified Relco SC480 glow starter allows, up to a resolution of $R \sim 30,000$, the calibration of the entire spectrum as well as of single individual orders.

8.1.9 The Feeding of the Calibration Light

Basically, there are three different ways to get the calibration light on the slit of the spectrograph.

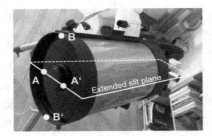

Figure 8.1 Position of the calibration light source (C8)

8.1.9.1 Feeding Inside the Spectrograph

This solution is implemented, for example, in Lhires III, Spectra L200 and SX spectrographs. In order to avoid a supply with a hazardous mains voltage usually a voltage converter, for example, 12 V/230 V is installed.

The SQUES echelle spectrograph allows the direct calibration light feeding of any source in front of the slit, for example by a Relco SC480 unit, via a separate fiber optic input.

8.1.9.2 Feeding Immediately in Front of the Spectrograph

For spectrographs, like DADOS, lacking of a direct feed for the calibration light, in Appendix D a proposal for the feeding through a hole in the bottom of a flip mirror is presented.

8.1.9.3 Feeding in Front of the Telescope

This type of feeding is considered a makeshift solution. Here the calibration light does not fall directly as a spherical wave on the slit, but it is fed via the convergent optical path of the telescope. With refractors the calibration light source must be strictly installed in the center of the front lens. With Cassegrain systems (Figure 8.1), the best position is as close as possible to the secondary mirror and in the extended plane, spanned by the longitudinal axis of the spectrograph slit and the optical axis of the telescope, for instance on the line between A and A'. The most unfavorable points are farthest away, at points B and B'. Tests have shown that here, a calibration error of >1 Å may result! Ideally, the optical path of the light source corresponds exactly to that of the examined star. For the determination of the calibration error see Section 11.1.6.

8.2 Calibration of the Spectral Flux Density (Intensity)

8.2.1 Preliminary Remarks

Compared to the wavelength the following explained calibration and normalization procedures for the spectral flux

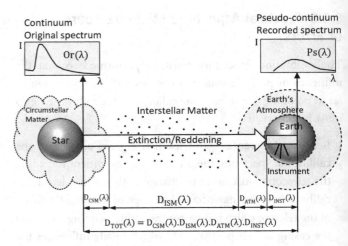

Figure 8.2 Continuum deformation of the original spectrum $Or(\lambda)$ by damping influences $D(\lambda)$

density (see Section 9.1.2) are much more complex. What kind of information we can extract from a spectrum depends strongly on how the intensity of the recorded profile was calibrated and normalized. As a result of an observation or measurement, such profiles are hereinafter denoted as "empirical functions" and abbreviated listed as $f(\lambda)$. This way, to each wavelength λ in the diagram a measured intensity value is assigned. Calculations with empirical functions are supported by the analysis software (e.g. Vspec).

8.2.2 Selective Attenuation of the Continuum Intensity

The intensity course of the undisturbed original stellar spectrum $Or(\lambda)$ is mainly determined by the blackbody radiation characteristics $B_{Teff}(\lambda)$ of a star (Section 1.4.2). On the long way to the recorded, unprocessed raw spectrum the continuum of $Or(\lambda)$ becomes deformed by the following damping influences $D(\lambda)$ into a hereinafter called pseudo-continuum $Ps(\lambda)$. Figure 8.2 schematically shows the deformation of the recorded profile as a result of scattering-effects and instrumental influences.

8.2.2.1 Attenuation (or Extinction) by Circumstellar Matter, $D_{CSM}(\lambda)$

In some rather rare cases, a significant attenuation already takes place in the vicinity of the star by a scattering effect due to circumstellar dust and gas cloud. Textbook examples are the young T-Tauri objects [1]. The circumstellar attenuation may be here so large that the intensity deviation of the H-emission lines from the theoretical Balmer decrement (see Section 14.1.3) is even applied as a classification criterion [1]. Further the dense dust clouds around the Mira variables on the AGB may even generate an inverse Balmer decrement!

8.2.2.2 Attenuation by the Interstellar Matter, $D_{ISM}(\lambda)$

The main cause here is also the same scattering effect due to interstellar dust grains and gas. Moreover, the intensity is selectively much stronger dampened in the blue short wave part of the spectrum. Thus the maximum of the continuum radiation is apparently shifted in to the red, long-wavelength range, what is called "interstellar reddening." This must never be confused with the "redshift" of the wavelength. The extent of this effect follows the extinction law and depends on the distance to the object, the direction of the line of sight and is, not surprisingly, most intensive within the galactic plane. It can just roughly be estimated with a corresponding 3D model by F. Arenou *et al.* [32], [51].

8.2.2.3 Attenuation by the Earth's Atmosphere, $D_{ATM}(\lambda)$

This acts similarly to the interstellar reddening. Well-known effects are the reddish sunsets. The modeling of the atmospheric extinction is mainly applied in the professional sector. It depends inter alia on the zenith-distance z of the observed object, the altitude of the observation site and the meteorological conditions [33]. These parameters also modify the optical path length through the Earth's atmosphere, the so called "airmass."

8.2.2.4 Attenuation by Instrumental Responses $D_{INST}(\lambda)$

At the very end of the transmission chain this is caused by the system telescope–spectrograph–camera, including the image sensor and the numerous optical and electronic elements. Typically, for most of the currently used amateur cameras, the blue-violet range of the spectrum gets massively dampened. The instrument response can be determined quite precisely, for example by the exactly known radiation characteristic of a special calibration light source [6], [35], [36].

The resulting overall attenuating effect is:

$$D_{Tot}(\lambda) = D_{CSM}(\lambda)D_{ISM}(\lambda)D_{ATM}(\lambda)D_{INST}(\lambda) \quad \{8.1\}$$

The attenuating effect, $D_{Tot}(\lambda)$, might also be considered as an empirical "attenuation function." Theoretically it provides any wavelength λ with the correction factor between the continuum intensity of $Ps(\lambda)$ and $Or(\lambda)$.

$$D_{Tot}(\lambda) = \frac{Ps(\lambda)}{Or(\lambda)} \quad \{8.2\}$$

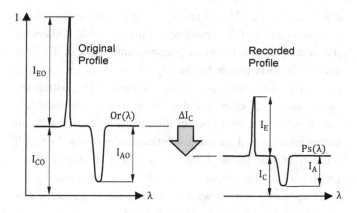

Figure 8.3 Proportional attenuation of the individual emissions and absorptions in relation to the continuum level

Unfortunately $D_{Tot}(\lambda)$ can be determined just by a very rough approximation. The intensity profile of the original stellar spectrum $Or(\lambda)$ can be simulated only on a theoretical basis and the individual attenuating factors, with exception of $D_{INST}(\lambda)$, can be estimated just very roughly.

8.2.3 Proportional Attenuation of the Spectral Lines

Figure 8.3 shows this proportional attenuating effect at an absorption and emission line. For the absorptions the attenuated peak intensities I_P are designated with I_A and for emissions with I_E; those in the undamped original profile with I_{AO} and I_{EO}. For the continuum intensity I_{CO} and I_C are applied. Within this area the continuum level is assumed to be running horizontally and in the recorded profile it appears to be lowered by the amount ΔI_C. The absorption lines, inseparably embedded into the continuum, are attenuated proportionally to the local continuum level $I_C(\lambda)$:

$$I_C = I_{CO}\, D_{Tot}(\lambda) \quad \{8.3\}$$

$$I_A = I_{AO}\, D_{Tot}(\lambda) \quad \{8.4\}$$

$$I_E \approx I_{EO}\, D_{Tot}(\lambda) \quad \{8.5\}$$

However, for the independently superimposed emission lines, the proportional attenuation of the peak intensity I_E must be just considered as a rough approximation. In fact, their entire energy fluxes F_E get dampened here separately, according to the extinction law.

8.2.4 Information Content of the Pseudo-Continuum

If the attenuation function $D_{Tot}(\lambda)$ would roughly be known, the pseudo-continuum of $Ps(\lambda)$ theoretically contains the information, to enable an approximate reconstruction of the original profile Or(λ), according to Equation {8.2}. Otherwise the course of the recorded pseudo-continuum of Ps(λ) is useless. It corresponds to the digital numbers representing the output of the analog-to-digital converter ADC, within the image sensor. It does not directly represent the recorded, wavelength-dependent photon flux, which is loaded with all the aforementioned attenuating influences. Anyway the rough relation of this incident flux to the "analog–digital units" (ADU) [6] at the output of the ADC is determined by the specific, mostly known, wavelength-dependent quantum efficiency QE (for definition see Section 6.2).

8.2.5 Proportional Procedures for the Relative Flux Calibration

In the following sections several calibration procedures and the associated effects are presented, which are also applied by amateurs. All are based on the division of Ps(λ) by empirical correction functions. Adapted to the specific application they scale the intensity of the pseudo-continuum with the embedded absorptions proportionally to an appropriate new level. For the independent emission lines the proportional scaling of the peak intensity I_E is just a rough simplification. Figure 8.4 shows how the absorptions in Ps(λ) are scaled on the level of Or(λ).

By this linear scaling the continuum-related measurement categories – peak intensity P and equivalent width EW – remain unchanged (see Section 9.1). They cannot proportionally be changed or scaled by a simple multiplication or division:

$$P_{\text{Absorption}} = \frac{I_{\text{AO}}}{I_{\text{CO}}} = \frac{I_A}{I_C} \qquad \{8.6\}$$

$$\text{EW}_{\text{Or}(\lambda)} = \text{EW}_{\text{Ps}(\lambda)} \qquad \{8.7\}$$

The I_A-values, measured in arbitrary units, appear to be scaled to I_{AO}. The original ratio of $I_{\text{AO}}/I_{\text{CO}}$ remains

unchanged in the pseudo-continuum {8.6}. For independently superimposed emission lines I_E, this proportional scaling of the peak intensity I_E is just a rough approximation. For higher requirements and objects, where the H Balmer series appear in emission, the energy flux of the individual emissions F_E can numerically be scaled to the ratios of the theoretical Balmer decrement (see Section 14.1.8).

8.2.6 Rectification of the Continuum Intensity

The most simple of the relative calibration processes is the rectification of the recorded profile Ps(λ), which is divided by its own "smoothed" or fitted intensity course Ps$_{\text{Fit}}$(λ). This way the continuum level runs horizontally (I_C = const.). The resulting profile is called "residual intensity" Ri(λ)[37]:

$$\text{Ri}(\lambda) = \frac{\text{Ps}(\lambda)}{\text{Ps}_{\text{Fit}}(\lambda)} \qquad \{8.8\}$$

By this division, the intensities of all spectral lines are proportionally scaled to a unified continuum level, mostly normalized to $I_C = 1$. This corresponds now approximately to the profile of a virtual star, showing a horizontally running and for all wavelengths uniform continuum radiation, what of course physically would be impossible. This method is considered as appropriate even for most professional applications [6].

Consequences and benefits of this procedure:

– The rectified profile Ri(λ) allows the elimination of the wavelength-dependent distribution of the stellar radiation intensity. It generates a "quasi neutralization," but never a real correction of the attenuating effects.
– Ri(λ) enables the direct visual comparison of the peak intensities between individual absorption lines, according

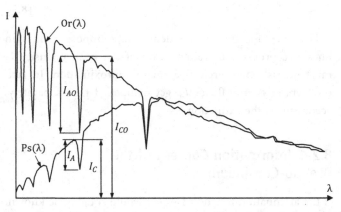

Figure 8.4 Scaling of the absorptions in Ps(λ) on the level of Or(λ)

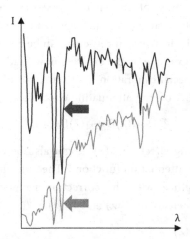

Figure 8.5 Scaling of the H and K absorptions

to $P = I_P/I_C$. The scaling effect of a rectified profile is impressively demonstrated by the Fraunhofer H and K absorptions (Ca II) in the solar spectrum (Figure 8.5). In the rectified continuum, the H and K lines appear now as the obviously strongest absorptions, which the Sun generates itself (black arrow).

– The unified radiation intensity of Ri(λ) allows the comparison of the emission line intensities with those of the theoretical, undamped Balmer decrement (see Section 14.1.3).

– The rectified profile, normalized to $I_C = 1$, enables the determination of the EW-values and facilitates the measurement of the FWHM and the Doppler shift $\Delta\lambda$.

8.2.7 Relative Flux Calibration by a Synthetic Continuum

The goal of this procedure is a rough approximation of the recorded profile Ps(λ), to the original and therefore not reddened continuum of Or(λ). The intensities of all spectral lines are here simply scaled to the fitted continuum course $Ms_{Fit}(\lambda)$, of a virtual model star of the same spectral class. Figure 8.6 shows the recorded pseudo-continuum Ps(λ) of Sirius. The upper profile is the aimed, fitted continuum course $Ms_{Fit}(\lambda)$ of the model star (VSpec library, CDS Database). It appears cleaned from all spectral lines and corresponds roughly to the blackbody radiation characteristics $B_{Teff}(\lambda)$ of an A1V star:

$$Ms_{Fit}(\lambda) \approx B_{Teff}(\lambda) \qquad \{8.9\}$$

In Figure 8.7, the dashed correction curve Ir(λ) is generated with a division of the fitted pseudo-continuum $Ps_{Fit}(\lambda)$ by the synthetic reference profile $Ms_{Fit}(\lambda)$ as displayed in Figure 8.6. It is denoted hereinafter with correction function Ir(λ):

$$Ir(\lambda) = Ps_{Fit}(\lambda)/Ms_{Fit}(\lambda) \qquad \{8.10\}$$

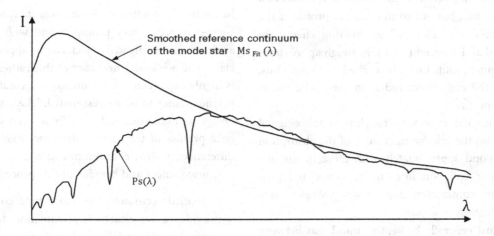

Figure 8.6 Recorded profile Ps(λ) of Sirius and aimed reference continuum $Ms_{Fit}(\lambda)$

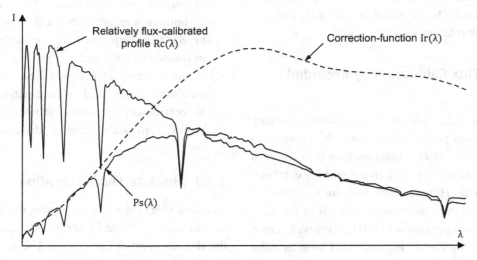

Figure 8.7 Correction curve Ir(λ) and relatively flux calibrated profile Rc(λ)

Note: In some papers, for example by ESO [34], $Ir(\lambda)$ is introduced and accordingly applied reciprocally as $Ms_{Fit}(\lambda)/Ps_{Fit}(\lambda)$.

In a rough approximation $Ir(\lambda)$ corresponds here to the "attenuation function" $D_{Tot}(\lambda)$, according to Equations {8.1} and {8.2}:

$$Ir(\lambda) \approx D_{Tot}(\lambda) \qquad \{8.11\}$$

Finally the division of the recorded profile $Ps(\lambda)$ by the correction-function $Ir(\lambda)$, results in the relatively flux-calibrated profile $Rc(\lambda)$. It shows now the same continuum course like the smoothed reference profile $Ms_{Fit}(\lambda)$ in Figure 8.6, but appears now overprinted with the accordingly scaled lines of the recorded profile:

$$Rc(\lambda) = Ps(\lambda)/Ir(\lambda) \qquad \{8.12\}$$

Consequences and benefits of this procedure:

– With the transformation of the intensity profile from $Ps(\lambda)$ to $Ms(\lambda)$, the recorded profile is just scaled to the level of a synthetic model star, but not to the original profile of the observed object $Or(\lambda)$. Thus, all attenuating effects are simply bypassed and a straight and very rough approximation to the original profile $Or(\lambda)$ is reached this way. Thus, fluctuations of the continuum radiation cannot be measured in such a profile.

– Since $Rc(\lambda)$ complies now very roughly to the original profile $Or(\lambda)$, also the relative intensities of the absorption lines I_A correspond approximately with those in the original profile. For emission lines I_E, generated independently from the continuum, this applies just as a very rough approximation.

– However, it must generally be kept in mind that between different stars, even of the very same spectral class, considerable differences in the continuum course may occur. This effect can significantly be enhanced by a strongly different metallicity and/or rotation velocity ($v \sin i$).

8.2.8 Relative Flux Calibration by Recorded Standard Stars

Similarly to Section 8.2.7, with this fairly time consuming procedure the recorded profile $Ps(\lambda)$ is scaled by a correction function $Ir(\lambda)$. However, $Ir(\lambda)$ is obtained here by a recorded and real existing standard star $St(\lambda)$. In the relevant databases they are labeled with HD numbers and are for practical reasons (e.g. fitting of the continuum) preferably of the spectral type A. The continuum course of $St(\lambda)$ is very well known and corresponds to the profile, just reddened by interstellar

matter, as it would have been recorded outside the Earth's atmosphere and without any instrumental responses. Such flux-calibrated spectra, compiled from different sources can be found, for example, in the ISIS software [71].

Standard stars must be recorded with a minimum of time difference and as close as possible to the investigated object. Subsequently, the fit to the obtained raw profile is divided by the fitted continuum course of the specific reference spectrum of the very same star from the catalog. Thus, in one step, the atmospheric $D_{ATM}(\lambda)$ and instrumental influences $D_{INST}(\lambda)$ can be corrected in a good approximation. Anyway, the resulting spectrum remains here reddened by the interstellar matter. The extent of reddening depends on the distance and the specific line of sight to the star and is in the "close range" of a few dozen light years, just very slight [6], [51]. In contrast to Equations {8.1} and {8.11} the correction function $Ir(\lambda)_{Standardstar}$ is determined here only by $D_{INST}(\lambda)$ and $D_{ATM}(\lambda)$:

$$Ir(\lambda)_{Standardstar} = D_{INST}(\lambda)D_{ATM}(\lambda) \qquad \{8.13\}$$

In contrast to Section 8.2.7 such real standard star correction curves, recorded very promptly and with similar elevation angle to the investigated object, can be applied to any spectral class. The achievable accuracy of this rather delicate method is highly dependent on the quality of execution. By amateurs it tends rather to be overestimated. The potential sources of errors are numerous and in addition even some of the reference profiles of the various databases may show significant differences in their continuum courses.

Consequences and benefits of this procedure:

– If applied accurately, this fairly time consuming method provides a reasonable approximation to the theoretical original profile $Or(\lambda)$, which still appears to be star specifically reddened by the interstellar matter.

– This way, at least theoretically, also greater fluctuations in the continuum radiation might be detectable.

– The effective temperature T_{eff}, according to Section 10.2, can roughly be estimated.

– For the intensity of emission lines even this correction procedure is just a rough approximation.

– This correction procedure is further required if spectra are obtained from certain scientific databases.

8.2.9 Absolute Flux Calibration

As a final step of the flux calibration, the intensity axis might be calibrated in physical units, mostly [erg cm^{-2} s^{-1} Å^{-1}], for the absolute spectral flux density. The absolute energy fluxes of the individual lines are usually calculated applying Equations {9.1} and {9.2} mostly in the [erg cm^{-2} s^{-1}]. This process is also based on the comparison of the absolute calibrated radiation flux of a standard star. A real absolute flux calibration is very challenging, time consuming and only needed by some special sectors of professional astronomy. However, many additional data are required here, such as the exposure time of the spectral recordings. Further such recordings must take place under quite rare "photometric conditions" and require a large slit width to measure the total flux of the object. Absolute flux calibration is relatively common for spectra, recorded by space telescopes, which of course remain clean of any atmospheric influences. For amateur applications an acceptable accuracy of the results would be prevented by the inadequate quality of the observation site. Even in the professional sector, such absolutely flux-calibrated spectra can be found rather rarely.

8.2.10 Tasks and Required Calibration Procedures

Table 8.1 provides a list of tasks and an overview of which calibration, normalization or correction procedures are appropriate or required.

CHAPTER

9 Analysis of the Spectra

9.1 Measurement of Spectral Lines

9.1.1 Measurement of the Wavelength in a Spectral Profile

On an amateur level in an accordingly calibrated spectrum the wavelength of a spectral line, mostly expressed in nanometer [nm] or ångstrom [Å], can be relatively easily obtained by the cursor position of the analysis software at the peak of the line or by the wavelength at the barycenter of a Gaussian fit (e.g. VSpec). Which method is preferable depends upon whether a strongly asymmetric blend or an isolated single line needs to be analyzed.

9.1.2 Intensity Measurement in a Spectral Profile

Much more demanding and complex are calibration and measurement in the context of the intensity axis. The intensity of the continuum and the spectral lines is defined as "spectral flux density" $I(\lambda)$, hereinafter also referred to as "intensity" I. In the astronomical literature F_λ, F_ν, f_λ, f_ν are also in use. Spectral flux density describes the rate of energy transferred by photons or electromagnetic radiation at a surface per unit area and unit wavelength, mostly expressed in [erg cm^{-2} s^{-1} Å^{-1}]. In the wavelength domain of radio astronomy this unit is called jansky (Jy) and defined as 10^{-23} erg cm^{-2} s^{-1} Hz^{-1} or 10^{-26} W m^{-2} Hz^{-1}.

Depending on the task the line intensity is determined either very complexly with an absolute calibrated scale but mostly by a simple relative measurement. Here, the latter method is presented which in the majority of cases is sufficient, for example for the intensity comparison between spectral lines. As a relative reference for the intensity either the

local continuum level $I_C(\lambda)$ is applied or sometimes a linear but otherwise arbitrary scale of the intensity axis.

9.1.3 Peak Intensity I_P and Energy Flux F of a Spectral Line

The peak intensity I_P and the energy flux F are the two most important measures of a spectral line, which complement each other. Figure 9.1 shows the continuum level $I_C(\lambda)$ displayed simplified as a horizontal black line. It is superimposed by an emission line, and lowered by an absorption line.

9.1.4 Peak Intensity and Energy Flux of an Absorption Line

The absorption line with the intensity $I_A(\lambda)$ and the peak intensity I_P, causes a drop in the local continuum at the

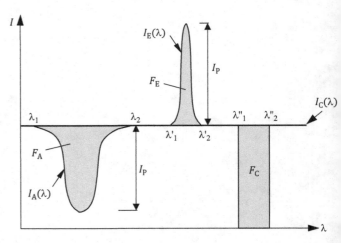

Figure 9.1 Continuum level $I_C(\lambda)$, with an emission and absorption line

element-specific wavelength λ of defined area, shape and line depth. It may simplified be considered as the product of a "filtering process." Therefore, all parameters of the absorption lines always remain inseparable and proportionally connected to the continuum intensity $I_C(\lambda)$. With stellar objects in most of the cases the lacking photons get absorbed in the photosphere. The peak intensity I_P corresponds here to the difference between the minimum spectral flux density $I_A(\lambda)_{min}$ at the lowest point of the absorption line, and the local continuum level $I_C(\lambda)$. The total energy flux F_A, being removed from the local continuum radiation, corresponds to the area of the absorption line, measured below the continuum level. Mathematically, this area in the range between λ_1 and λ_2 is expressed as the sum ($\int$) of an infinitely large number of rectangles with the heights $I_C(\lambda) - I_A(\lambda)$ and the infinitely small width $d\lambda$:

$$F_A = \int_{\lambda_1}^{\lambda_2} [I_C(\lambda) - I_A(\lambda)]d\lambda \qquad \{9.1\}$$

With the absolute calibrated intensity scale for F_A usually the unit [erg cm^{-2} s^{-1}] is applied.

9.1.5 Peak Intensity and Energy Flux of an Emission Line

The emission line with the intensity $I_E(\lambda)$ and the peak intensity I_P is generated independent of the continuum, in excited atoms by electron transitions downwards to lower levels. It is therefore, neither with regard to its shape nor its intensity, in any way connected with the continuum.

The continuum level $I_C(\lambda)$, however, is dependent on the specific spectral flux density the star is generating at the wavelength λ. Therefore the emission line appears fully independently superimposed on the local continuum level [22]. This results in the superposition of both intensities. Within the range of the emission line the resulting profile $I_E(\lambda)$ contains additionally the local continuum level $I_C(\lambda)$.

Due to the physically, and mostly locally, different generation, $I_C(\lambda)$ and $I_E(\lambda)$ may fluctuate independently of each other. It results in a certain strongly object-dependent and time related degree of coupling between the intensities of the continuum radiation and the emission lines. For instance P Cygni generates these lines immediately in its turbulent expanding envelope and Be stars in the somewhat distant circumstellar gas disk[1]. In the H II regions or planetary nebulae PN, this process takes place even up to some light years away from the ionizing star, where almost "laboratory conditions" exist!

The peak intensity I_P corresponds here to the difference between the maximum spectral flux density $I_E(\lambda)_{max}$ in the upper apex of the emission line and the local continuum level $I_C(\lambda)$. The total energy flux F_E corresponds to the area of the emission line, measured above the continuum level $I_C(\lambda)$ [37]. If it really appears superimposed to a continuum, as displayed in Figure 9.1, it must therefore be subtracted from the local continuum flux. Thereby, it becomes negative because $I_E(\lambda) > I_C(\lambda)$, which is in accordance with the convention (see also Section 9.1.9):

$$F_E = \int_{\lambda_1'}^{\lambda_2'} [I_C(\lambda) - I_E(\lambda)]d\lambda \qquad \{9.2\}$$

With the absolute calibrated intensity scale for F_E usually the unit [erg cm^{-2} s^{-1}] is applied.

9.1.6 Energy Flux of the Continuum

Simplified as a gray rectangle in Figure 9.1 an arbitrary area is displayed, where in a range between λ_1'' and λ_2'', the energy flux F_C of the continuum radiation is measured. Here, F_C corresponds to the area of the rectangle:

$$F_C = \int_{\lambda_1''}^{\lambda_2''} I_C(\lambda)d\lambda \qquad \{9.3\}$$

With the absolute calibrated intensity scale for F_C usually the unit [erg cm^{-2} s^{-1}] is applied.

9.1.7 Superposition of Emission and Absorption Lines

The superposition of emission and absorption lines results in a superposition of both intensities (see Figure 9.2). Within the range of the emission line the resulting profile $I_E(\lambda)$ contains now additionally the level $I_A(\lambda)$ of the absorption line.

For example, in Be-stars, the slim hydrogen emission line is generated in the circumstellar disk or shell, and appears superimposed to the rotational- and pressure-broadened H-absorption of the stellar photosphere. The resulting spectral feature is therefore called the "shell core" or "emission core" [7]. For a typical example refer to δ Sco [1]. The H absorption of such a spectral feature may also originate from the photosphere of a hot O star and the emission line from the surrounding H II region; see, for example, the Hβ line of θ^1Ori C [1].

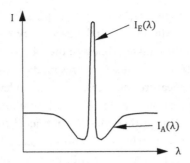

Figure 9.2 Superposition of an emission and an absorption line

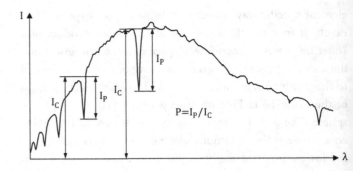

Figure 9.3 The continuum-related peak intensity P

9.1.8 The Continuum-Related Peak Intensity P

The easiest way is the direct measurement of the peak intensity I_P by a linearly but otherwise arbitrarily scaled intensity axis. However, this measure is only significant in a relatively or even absolutely flux-calibrated profile according to Section 8.2. Thus, for a relative measurement in many cases the peak intensity I_P is referred to the local continuum level $I_C(\lambda)$. Only this way, in a non-intensity-calibrated profile, I_P gets comparable with other lines. This results in the dimensionless continuum-related peak intensity P (see Figure 9.3).

$$P = I_P/I_C \qquad \{9.4\}$$

In the case of absorption lines, in literature I_P is sometimes alternatively denoted as "LD" for "line depth" and P as "LDR" for "line depth ratio." If emission lines appear superimposed on a continuum for the observation of individual lines, sometimes, as a pure makeshift, their intensity I_P is referred to as the possibly independently fluctuating level $I_C(\lambda)$.

9.1.9 The Continuum-Related Energy Flux: Equivalent Width EW

In many cases for a relative measurement also the energy flux F of a spectral line is referred to the local continuum level $I_C(\lambda)$. This is accomplished by the so-called "equivalent width," or "EW" value. It is a relative and dimensionless measure for the area or the energy flux of a spectral line.

9.1.9.1 Definition

The profile area between the rectified continuum level, normalized to unity $I_C = 1$ and the profile of the spectral line, has the same size as the rectangular area with the depth $I_C = 1$ and the equivalent width EW [Å] (see Figure 9.4):

$$\text{EW} = \frac{\text{Profile area}}{I_C} \qquad \{9.5\}$$

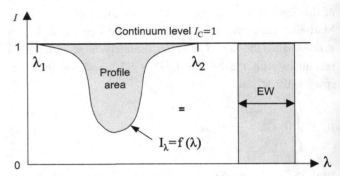

Figure 9.4 Definition of the equivalent width EW

Generally formulate, and according to Equation $\{9.1\}$ or $\{9.2\}$, the energy flux of the line, F_A or F_E, must now be divided by the local continuum flux $I_C(\lambda)$. In this case, theoretically a normalization of the continuum to unity would not be necessary. However, as a challenging requirement, the course of the quasi-continuum $I_C(\lambda)$, between λ_1 and λ_2 would have to be known.

According to Equation $\{9.1\}$ for an absorption line this results in:

$$\text{EW}_A = \int_{\lambda_1}^{\lambda_2} \frac{I_C(\lambda) - I_A(\lambda)}{I_C(\lambda)} \, d\lambda \qquad \{9.6\}$$

According to Equation $\{9.2\}$ for an emission line this results in:

$$\text{EW}_E = \int_{\lambda_1'}^{\lambda_2'} \frac{I_C(\lambda) - I_E(\lambda)}{I_C(\lambda)} \, d\lambda \qquad \{9.7\}$$

According to convention for an emission line the EW value is negative, due to $I_E(\lambda) > I_C(\lambda)$.

Notes to EW and FWHM values:

In scientific publications EW is sometimes labeled with W. For example, W_α designates the equivalent width of the

Hα line. Sometimes even the minus sign of the EW of emission lines is omitted. It is somewhat confusing that in some publications the FWHM value is also expressed as W. The conclusion: One must always simply check which value is really meant.

9.1.10 Normalized Equivalent Width W_λ

Rather rarely the normalized EW value W_λ is applied:

$$W_\lambda = \frac{EW}{\lambda} \qquad \{9.8\}$$

This allows the comparison of EW-values of different lines at different wavelengths λ, taking into consideration the linearly increasing photon energy towards decreasing wavelength λ or increasing frequency ν, according to Equation {1.2}.

9.1.11 Full Width at Half Maximum Height (FWHM)

The FWHM value is the line width in [Å] at half the height of the maximum intensity. It can be correctly measured even in non-normalized spectral profiles. Among other conditions, the width of a spectral line depends on temperature, pressure, density and turbulence effects in stellar atmospheres. It therefore allows important conclusions and is often used as a variable in empirical equations, for example to determine the rotational velocity of stars (see Section 12.1.7).

9.1.11.1 FWHM as Wavelength Difference

In most cases FWHM is specified as wavelength difference $\Delta\lambda$ in [Å], as displayed in Figure 9.5.

9.1.11.2 FWHM as Doppler Velocity

In the context of the measurement of rotational and expansion velocities, FWHM is also expressed as a velocity value according to the Doppler law. For this purpose FWHM [Å] is inserted, instead of $\Delta\lambda$, into the spectroscopic Doppler equation {11.1}. This way it gets converted into a velocity value [km/s]:

$$FWHM_{Doppler} = \frac{FWHM}{\lambda_0} c \qquad \{9.9\}$$

First of all the FWHM value, obtained from the spectral profile, has to be corrected from the instrumental broadening:

$$FWHM_{corr} = \sqrt{FWHM^2_{measured} - FWHM^2_{instrument}} \quad [Å] \qquad \{9.10\}$$

$FWHM_{Instrument}$ is determined by the resolution R of the spectrograph. It can normally be found in the spectrograph's manual as the R-value, which is valid for a defined wavelength range (λ = considered wavelength) [37]. In another context R was introduced in Section 4.2.3 and depends on the smallest possible dimension $\Delta\lambda$ [Å] of a line detail, which can just be resolved:

$$R = \lambda / \Delta\lambda, \qquad FWHM_{Instrument} = \frac{\lambda}{R} \qquad \{9.11\}$$

This value is determined by FWHM measurements at slimmest possible spectral lines, for example atmospheric H_2O absorptions or somewhat less accurate, at faint emission lines of calibration light sources [6], [37]. In laboratories, for example, emission lines generated by microwave excited mercury lamps are used in order to minimize the temperature broadening. The theoretical slimmest possible profiles are called "instrumental profile" or "δ-function response" [6]. The resolution may further be limited by a too rough pixel grid of the connected camera.

9.1.12 Half Width at Half Depth (HWHD or HWHM)

Very rarely for certain applications the half of the FWHM value is used, called HWHD and sometimes abbreviated as HWHM (half width at half depth or half maximum height):

$$HWHD = \frac{FWHM}{2} \qquad \{9.12\}$$

9.1.13 Full Width at Zero Intensity (FWZI)

Sometimes the FWZI value of a spectral line is applied. The "full line width at zero intensity" corresponds to the broadest part of the spectral line or to the integration range $\lambda_2 - \lambda_1$ of the definite integral according to Figure 9.1:

$$FWZI = \lambda_2 - \lambda_1 \qquad \{9.13\}$$

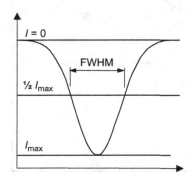

Figure 9.5 Definition of the FWHM

9.1.14 Half Width at Zero Intensity (HWZI)

The half of FWZI, the HWZI value, or the "half line width at zero intensity" is sometimes required for a detailed analysis of a single spectral line:

$$\text{HWZI} = \frac{\text{FWZI}}{2} \qquad \{9.14\}$$

Expressed as a velocity value according to the Doppler principle (Equation {9.9}), HWZI is applied instead of FWHM, for the estimation of the expansion velocities in nova spectra [1].

9.1.15 Measurement of Asymmetry

The asymmetry of a spectral line can be measured either by mirroring or by its so called "bisector line" (Figure 9.6)[6]. The wavelength axis has to be calibrated here according to the

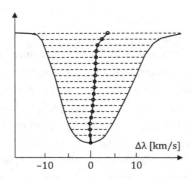

Figure 9.6 Bisector line of an asymmetric absorption

Doppler velocity $v = (\Delta\lambda/\lambda_0)c$. One of the possible applications is the study of stellar granulation cells [6].

9.1.16 Influence of the Spectrograph Resolution on the FWHM and EW Values

The spectral profiles of the Sun in Figure 9.7, recorded with different high resolutions show the influence on the recorded spectral lines. The R-values are here within a range of approximately 800–20,000.

The comparison of these profiles shows the following:

– If the resolution R is increased it becomes clearly evident that in stellar profiles, particularly of the middle to late spectral classes, practically no "pure" lines exist. Apparent single lines almost turn out as a "blend" of several sublines, if considered at higher resolutions.

– Striking is also the effect of the so-called "instrumental broadening." Even relatively well-insulated seeming lines broaden dramatically with decreasing resolution R, due to the instrumental influences. This affects the measured FWHM of a line (Section 9.1.11).

– The equivalent width EW of a line remains theoretically independent of the resolution. At higher resolutions, the area of the slimmer line profile normally gets compensated by a higher peak-value P. This is limited to very rarely appearing, discrete and well isolated single lines. So, for example, with a high-resolution spectrograph we can measure just one, well-defined single line. Otherwise at low resolution and the same wavelength, we would possibly

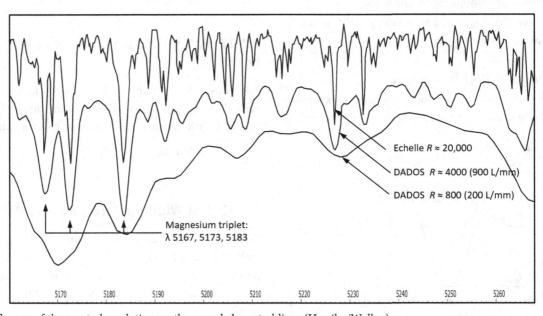

Figure 9.7 Influence of the spectral resolution on the recorded spectral lines (Huwiler/Walker)

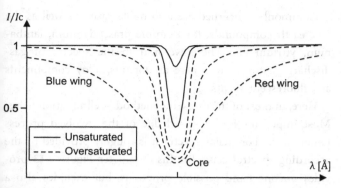

Figure 9.8 Absorption lines with different degrees of saturation (schematically after [6])

find a much larger EW value measured at a blend of several inseparable lines. Therefore such a measurement inevitably requires the declaration of the spectrograph resolution (*R*-value).

9.1.17 Additional Measurement Options

Depending on the applied analysis software, further information can be obtained from the calibrated spectral profile for example the SNR or the spectral dispersion in Å/pixel, etc.

9.2 Shape and Intensity of Spectral Lines

9.2.1 The Shape of Absorption Lines

Figure 9.8 shows, schematically after [6], several absorption lines with the same wavelength, but with different width and intensity. According to their degree of saturation, they penetrate differently, deep into the continuum. The profiles drawn in solid lines are unsaturated, showing an ideal, Gaussian-like intensity distribution. The absorptions displayed in dashed lines are oversaturated and appear massively broadened, without penetrating much further into the continuum. In astronomical spectra, an absorption line reaches full saturation before it touches the wavelength axis [6].

The lower part of the profile is called "core," which passes in the upper part over the "wings" into the continuum level. The short-wavelength wing is called "blue wing," the long-wave one "red wing" [13].

9.2.2 The Shape of Emission Lines

In contrast to the presented absorption lines, emissions always rise upwards from the continuum level. Oversaturated emission lines are very easy to recognize because they appear

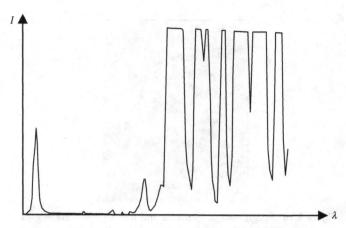

Figure 9.9 Overexposed Ne calibration spectrum, partly with oversaturated emission lines

flattened on the top. With the exception of two lines in the left part of the profile, Figure 9.9 shows an overexposed and therefore unusable calibration spectrum.

9.2.3 The Information Content of the Line Shape

There exist hardly any stellar spectral lines exhibiting an ideal shape. But a wealth of information about the object is hidden in the deviation from this. Here are some examples of physical processes which have a characteristic influence on the profile shape and become measurable this way:

– The rotational velocity of a star, caused by the Doppler effect, flattens and broadens the line (rotational broadening, Section 12.1.8).
– The temperature, density and pressure of the stellar atmosphere broaden the line (temperature/pressure/collision broadening).
– Macro turbulences in the stellar atmosphere broaden the line, caused by the Doppler effect.
– The combined effects of pressure and Doppler broadening result in so-called "Voigt profiles."
– Instrumental responses broaden the line (instrumental broadening, Section 9.1.11).
– In strong magnetic fields (e.g. sunspots) a splitting of the spectral line occurs due to the Zeeman effect (Section 13.2.2).
– Electric fields generate a similar phenomenon, the so-called Stark effect (Section 13.2.2).

9.2.4 Blends

Stellar spectral lines are usually more or less strongly deformed by closely neighboring lines, causing so-called

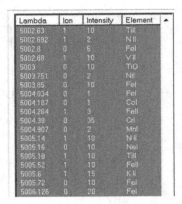

Lambda	Ion	Intensity	Element
5002.63	1	10	TiII
5002.692	1	2	N II
5002.8	0	6	FeI
5002.88	1	10	V II
5003	0	10	TiO
5003.751	0	2	NiI
5003.85	0	10	FeI
5004.034	0	1	FeI
5004.187	0	1	CoI
5004.264	1	3	FeII
5004.38	0	35	CrI
5004.907	0	2	MnI
5005.14	1	10	N II
5005.16	0	10	NeI
5005.18	1	10	TiI
5005.52	1	10	FeII
5005.6	1	15	K Ii
5005.72	0	10	FeI
5006.126	0	20	FeI

Figure 9.10 Assignment of elements and ions to rest-wavelengths (Vspec/Tools/elements/lineident)

"blends." The lower the resolution of the spectrograph, the more lines are merged to blends.

9.3 Identification of Spectral Lines

9.3.1 Task and Requirements

With the identification of a line to an absorption or emission with the wavelength λ the responsible element or ion is assigned. Considered purely theoretical this would be relatively easy, as shown in Figure 9.10 by the cut out of the "lineident" table, provided by the VSpec software. Here, unambiguously corresponding elements and ions are assigned to the rest wavelengths, λ_0. In practice, however, the following should be noted:

– Low noise spectra are required with a high SNR, which are calibrated very precisely by rest wavelength λ_0, and therefore free of any possible Doppler shifts.
– The higher the resolution of the spectrum, the more accurately λ can be determined and the fewer lines are merging into blends.

In the professional sector this process is widely supported and even automated by specific software.

9.3.2 Practical Problems and Strategies for Solving Them

Figure 9.10 shows that in certain sections of the spectrum, the distances between the individual positions are obviously very close. This happens, from quantum mechanical reasons, for a very large number of the metal lines, generating corresponding ambiguities, particularly in stellar spectra of the middle and later spectral classes.

Commonly concerned are also noble gases, as well as the rare earth compounds, for example praseodymium, lanthanum, yttrium etc. These we find in the spectra of gas-discharge lamps, acting here as dopants, alloy components and fluorescent agents [1].

Here, in most of the cases, the method of elimination helps. Most important is the knowledge of the involved process temperature. For stellar spectra it is roughly provided by the according spectral class. By this parameter Figure 9.13 provides on one hand possible proposals, but excludes also a priori certain elements or corresponding ionization stages.

At certain stages of stellar evolution, detailed knowledge of the involved processes are necessary. For example, since stars in the final Wolf Rayet stage first of all repel their entire outer hydrogen shell, this element can subsequently hardly be detected anymore in such spectra. Particularly critical in this case is the mostly very significant He II emission at $\lambda 6560.1$, which is often misinterpreted by amateurs as Hα line at $\lambda 6562.82$ [1].

Relatively easy is the line identification for calibration lamps with known gas filling. Thus, Vspec allows the superimposing of the corresponding emission lines, with their relative intensities, directly onto the wavelength-calibrated lamp spectrum. In cases of unknown gas filling, on a trial basis, the emission lines of the individual noble gases He, Ne, Ar, Kr and Xe can be superimposed on the calibrated lamp spectrum. In most cases the pattern of these inserted lines instantly shows if the corresponding element is present or not. This was also the most successful tactic for the line identification generated by the modified glow starter lamps [1]. However, some of the noble gas emissions can be located very close to each other such as Ar II at $\lambda 6114.92$ and Xe II at $\lambda 6115.08$.

9.3.3 Tools for the Identification of Spectral Lines

For stellar spectra, a spectral atlas is probably the safest way to identify spectral lines. For high-resolution echelle spectra of main sequence stars Spectroweb by A. Lobel is available [23]. For rare stellar types, object related publications are often very helpful. Appropriate software solutions, based on model spectra are primarily applied in the professional astronomy and are hardly suitable for most amateurs. For a detailed analysis of individual elements and their ions also online databases are available, such as from the National Institute of Standards and Technology (NIST) [27]. Figure 9.11 shows a Vspec screenshot with the

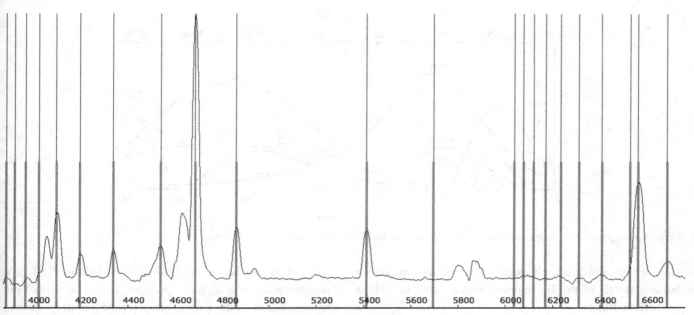

Figure 9.11 Calibrated DADOS spectrum of WR 136 with superimposed He II lines (VSpec)

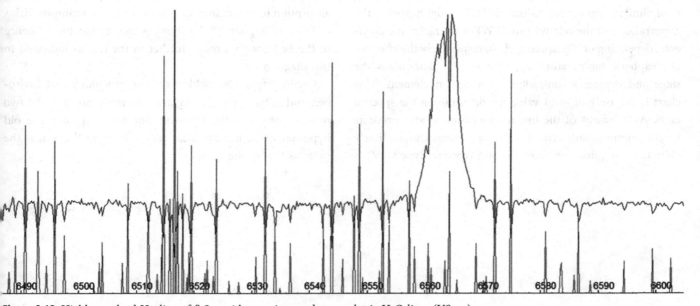

Figure 9.12 Highly resolved Hα line of δ Sco with superimposed atmospheric H_2O lines (VSpec)

calibrated DADOS spectrum of the Wolf Rayet star WR 136, superimposed by the He II emission lines from the "lineident" database. For comments on the spectrum refer to [1].

Figure 9.12 shows a VSpec screenshot with a wavelength-calibrated, high-resolution SQUES echelle spectrum of δ Scorpii around the Hα line. It is superimposed with the slim atmospheric water vapor lines (H_2O), generated by the "lineident" database function. For comments on the spectrum refer to [1].

9.4 Temperature Related Appearance of Elements and Molecules in the Spectra

After studies in Cambridge and Harvard, Cecilia Payne-Gaposchkin (1900–1979) was the very first to apply the laws of nuclear physics to astronomy. She made several pioneering discoveries, published, according to Otto Struve, in the "undoubtedly most brilliant Ph.D. thesis ever written in astronomy" [2]. First of all she showed that the spectrum of a star is primarily determined by the temperature of the

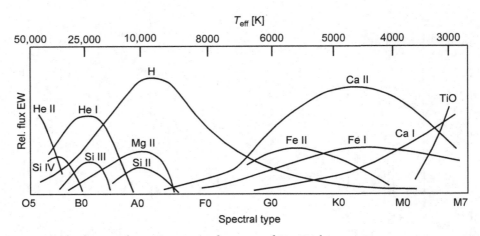

Figure 9.13 Approximate range of influence of some important elements and ions in the spectrum

photosphere, which directly defines the spectral class on the horizontal axis of the Hertzsprung–Russell diagram (HRD; see Section 10.1.1).

Further, Cecilia Payne-Gaposchkin provided the theoretical basis for the chart in Figure 9.13, which for spectroscopy is of similar importance to the HRD. It roughly shows the appearance and the relative flux (EW) of characteristic elements, depending on the spectral class, respectively the effective photospheric temperature T_{eff} of a star. It determines the stage and degree of ionization for a certain element. This chart is not only of great value for determining the spectral class. As a benefit of the line identification it also prevents severe interpretation errors. Thus, it becomes immediately clear that the photosphere of the Sun (spectral type G2V) is

a few thousand kelvin too cold to show helium He I in a normal (photospheric) solar spectrum. It also shows that the hydrogen lines of the Balmer series remain visible in varying intensities in nearly all spectral classes. Only in the late M classes, do they get increasingly overprinted by strong absorption bands, mainly caused by TiO. The examples Sirius (A1V) and Regulus (B7V) show, however, that the influence of the Fe lines goes much further to the left, as indicated in this diagram.

Cecilia Payne-Gaposchkin further concluded that hydrogen and helium, the two lightest elements, are also the two most common in the Universe. She also disproved the old hypothesis that the chemical composition of the Sun is the same as that of the Earth.

CHAPTER

10 Temperature and Luminosity

10.1 Information Content of the Spectral Classification

10.1.1 The Hertzsprung–Russell Diagram

The HRD was developed independently by Ejnar Hertzsprung in 1906 and by Henry Russell 1913. It is probably the most fundamental and powerful illustrating tool in astrophysics and of extraordinary importance for astrospectroscopy. On the topic of stellar evolution in the context of the HRD an extensive literature exists, which should be studied by the ambitious amateur. Here, just the information content of the spectral classification is demonstrated, if analyzed in the context of the HRD diagram.

10.1.2 Information Content

Figure 10.1 shows the HRD populated with stars, which are presented in the *Spectral Atlas* [1]. The Sun, along with Sirius, Vega, Regulus and Spica, are located as dwarf stars on the main sequence. Arcturus, Aldebaran, Pollux, etc. have left the main sequence, and are located on the giant branch shining with luminosity class III. Deneb, Rigel and Betelgeuse are in the realm of giants, with the respective class Ia–Ib. Down on the branch of the white dwarfs, the remnants of stars gather which were originally ≲ 8 solar masses. In addition to 40 Eridani B and Van Maanen 2, the companions of Sirius and Procyon are located here.

The absolute luminosities of the stars are plotted against the spectral type or the effective temperature respectively. However, the exact temperature boundaries between the classes may differ somewhat, depending on the consulted source. This concerns particularly the lower limit of the O class (25,000–29,000 K), and the late spectral types, which for each luminosity class may exhibit different temperature calibrations in the range of several hundred kelvin. For further details refer to Gray and Corbally [7].

The spectral class unambiguously determines the position of a star on the diagram. In Figure 10.1, the position of the Sun – shown with bold arrows – can be estimated from the individual scales of the diagram. On the upper edge of the diagram we find the effective temperature of about 5700 K. On the left we see the absolute magnitude of $M_V \approx 4.8$, corresponding to the apparent brightness, which a star generates in a normalized distance of 10 parsecs, or some 32.6 light years. Finally, the scale on the bottom indicates the spectral type of the Sun, G2.

10.1.3 Spectral Class, Stellar Mass and Life Expectancy

The effect of the initial stellar mass is crucial for stellar evolution. First, the initial stellar mass determines which place a star occupies on the main sequence – the larger the mass the more to the left on the HRD, or "earlier" in respect of the spectral classification. Second, it has a dramatic impact on the star's entire life expectancy, and the somewhat shorter period of time that it will spend on the main sequence. This ranges from roughly some million years for the early, extremely hot O types up to >100 billion years for the cool red dwarfs of the M class. The more massive the star, the more energy it has to invest in order to avoid a gravitational collapse. Accordingly, its energy supply is exhausted more quickly. Table 10.1 shows this correlation, together with other

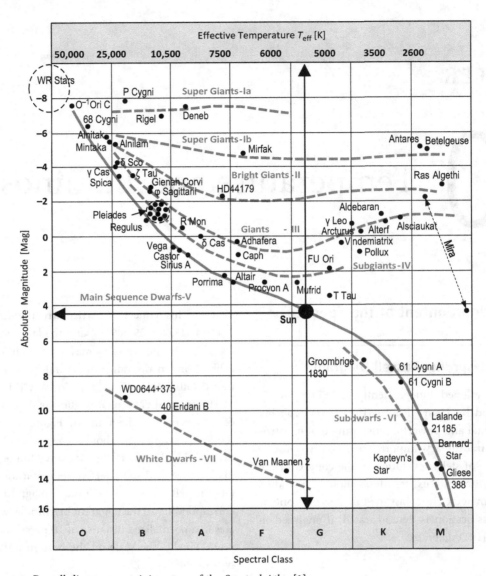

Figure 10.1 Hertzsprung–Russell diagram containing stars of the *Spectral Atlas* [1]

parameters, indicated for dwarf stars on the main sequence. The values provide a general overview and may also vary depending on source. Mass, radius and luminosity are indicated relative to the Sun ($\odot$).

10.1.4 The Evolution of the Sun in the HRD

The following simplified short description demonstrates how the spectral class provides a rough imagination of the state of stellar development. In Figure 10.2 the approximate evolutionary path of the Sun is displayed. By contraction of a gas and dust cloud, at first a protostar is formed, which subsequently moves within some million years onto the main sequence. Here, the luminosity stabilizes to about

70% of today's value. This point is called ZAMS for "zero age on main sequence." Within the next 9–10 billion years the luminosity increases to >180%, whereby the spectral class G2 remains more or less constant. During this period as a dwarf or main sequence star, hydrogen is fused into helium.

Towards the end of this stage the hydrogen is burning in a growing shell around the helium core. The star becomes unstable and expands to a red giant of the M class (luminosity class ~II). It moves on the red giant branch (RGB) to the top right of the HRD where after 12 billion years the ignition of the helium nucleus starts, the "helium flash." The photosphere of the red giant expands almost to the size of Earth's orbit. By this huge expansion the gravity

Table 10.1 Spectral class, stellar mass and expected life on the main sequence

Spectral Class Main Sequence	Mass $M/M_\odot$	Main Sequence Lifetime	Effective Temperature T_{eff} [K]	Radius $R/R_\odot$	Luminosity $L/L_\odot$
O	20–60	~1–10 My	≳25,000–50,000	9–15	90,000–800,000
B	3–18	~10–300 My	10,500–25,000	3.0–8.4	95–52,000
A	2–3	~300 My–2 Gy	7500–10,500	1.7–2.7	8–55
F	1.1–1.6	~2–7 Gy	6200–7500	1.2–1.6	2.0–6.5
G	0.9–1.05	~7–20 Gy	5000–6200	0.85–1.1	0.66–1.5
K	0.6–0.9	>20 Gy	3600–5000	0.65–0.80	0.10–0.42
M	0.08–0.6	>100 Gy	2600–3600	0.17–0.63	0.001–0.08

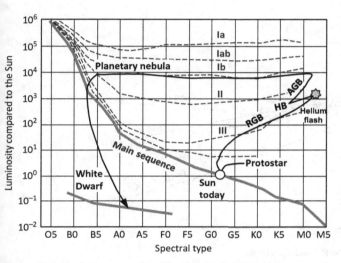

Figure 10.2 Evolutionary path of the Sun in the HRD

acceleration within the stellar photosphere is reduced dramatically and therefore the star, already at this stage, loses about 30% of its mass.

Subsequently the giant moves with fusing helium core on the horizontal branch (HB) to the intermediate stage of a yellow giant of the K class (lasting about 110 million years) and then via the reverse loop of the asymptotic giant branch (AGB) to the upper left edge of the HRD [1]. At this stage the star repels the remaining shell as a visible planetary nebula. Next, it shrinks to an extremely dense object of about Earth's size. After further cooling, it moves down on the branch of the white dwarfs. Later on the star will finally become invisible as a black dwarf, and disappear from the HRD. During its lifetime the Sun will pass through a large part of the spectral classes, but with very different luminosities.

10.1.5 The Evolution of Massive Stars

Due to highly complex nuclear processes such stars show in the giant stage complex pendulum-like movements in the upper part of the HRD, passing various stages of variables. During this period also many of the heavy elements up to iron are generated. For stars with $> 25\ M_\odot$ a Wolf Rayet and/or luminous blue variable (LBV) stage may follow [1]. Stars with an original mass of $>8\ M_\odot$ will not end as white dwarfs, but explode as supernovae. Dependent on the original mass the remnant will be a neutron star or in extreme cases a black hole.

10.2 Measurement of the Stellar Effective Temperature T_{eff}

10.2.1 Introduction

Depending on the spectral class, stellar spectra and their measurable features also reflect the physical state of the photosphere. For the spectroscopic determination of the effective temperature T_{eff} several, differently sophisticated methods exist. For a definition of T_{eff} refer to Section 1.4.4.

10.2.2 Temperature Estimation by the Spectral Class

The spectral class reflects directly the sequence of the corresponding effective temperatures. It is therefore the most simple but relatively imprecise way to estimate T_{eff}. In the literature numerous tables can be found, calibrating the effective temperatures relative to the individual spectral and luminosity classes. However, even between renowned sources, significant differences may arise. Further, the luminosity classes of the individual spectral types are calibrated with considerably

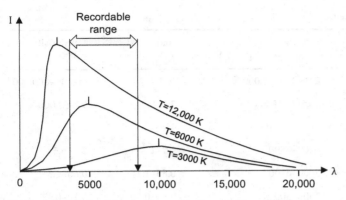

Figure 10.3 Maximal intensities and corresponding effective temperatures, related to the recordable spectral domain

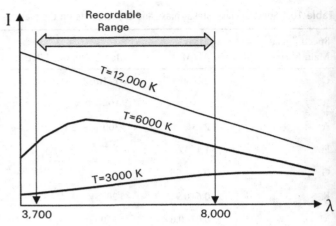

Figure 10.4 Different slopes $dI/d\lambda$ of the continua within the recordable spectral domain

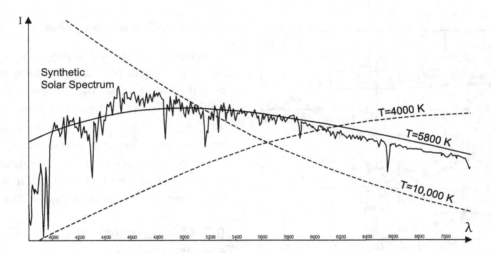

Figure 10.5 The solid curve representing $T_{eff} = 5800$ K fits to the synthetic solar spectrum

different temperatures. For main sequence stars of middle and late spectral classes, the accuracy, realistically achievable by amateurs, is within a few 100 K. Further it should be noted, that the determination of the luminosity class is rather difficult. In the *Spectral Atlas* [1] Appendix D shows T_{eff} values as a function of the spectral and luminosity class, based on the *Stellar Spectral Flux Library* by A. J. Pickles [109].

10.2.3 Temperature Estimation by Applying Wien's Displacement Law

A further possible approach is to estimate T_{eff} by the principle of Wien's displacement law (Section 1.4.3). It is based on the simplified assumption that the radiation characteristic of a star corresponds roughly to a blackbody. Theoretically, T_{eff} could be estimated by Equation {1.8} and based on the wavelength λ, which has been measured at the maximum intensity

of the profile. This requires, however, relatively flux-calibrated profiles according to Section 8.2.

Further, the maximum intensity must be located here within the recorded range – for a typical amateur spectrograph about 3800–8000 Å. In Figure 10.3 this criterion is met only by the graph, representing 6000 K. According to Equation {1.7}, within this section, only profiles with T_{eff} of about 3600–7600 K can be analyzed by their maximum intensity. This corresponds roughly to the spectral types M1–F0.

Therefore, the coverage of all spectral classes requires an adaptation of this method. One possibility is enabled by the function between the continuum slope $dI/d\lambda$ and T_{eff} (Figure 10.4).

This method is provided, for example, by the VSpec software, applying the function "Radiometry/Planck." Thereby the relatively flux-calibrated profile is iteratively fitted by displayed continuum curves, corresponding to certain T_{eff} values. This principle is demonstrated by Figure 10.5 with a

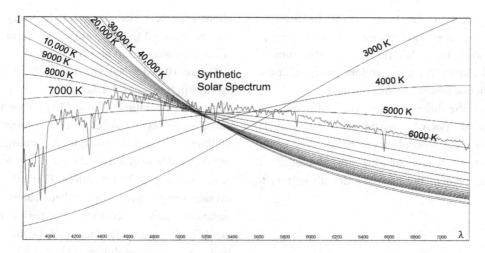

Figure 10.6 Array of curves with different continuum courses for the estimation of T_{eff} (Visual Spec Database)

synthetically generated solar spectrum G2V. The continuum curve is displayed, fitting to the Sun's profile with 5800 K. The maximum intensity of this curve is at ~λ4900, corresponding to the temperature according to Wien's displacement law. Further, displayed for comparison, are the curves for 10,000 K and 4000 K.

In practice, of course, the temperature estimation of the investigated star is requested, rather than the reproduction of an already known temperature of a synthetic profile. This requires a relative flux calibration of the obtained pseudo-continuum with a recorded standard star, according to Section 8.2.8. The accuracy of this measurement is directly dependent on the quality of this relatively demanding procedure.

Figure 10.6 shows, in front of the synthetic solar spectrum, the continua of various effective temperatures according to the VSpec tool "Radiometry/Planck." In the range between 3000 and 20,000 K, the increasingly closer following curves are displayed in 1000 K steps. Above 20,000 K the intermediate spaces become so close, that here only the continua for 30,000 and 40,000 K are shown. This explains why the temperature estimation is limited here to the middle and late spectral types. Further it gets obvious, that with increasing temperature, the differentiating influence of the spectral class to the flux calibration is significantly decreasing.

10.2.4 Temperature Determination Based on Individual Lines

Generally these methods are based on the temperature dependency of the line intensity $I = f(T)$. However, the intensity and shape of the spectral lines are determined by many other variables, such as element abundance, pressure, turbulence, metallicity Fe/H and the rotation speed of the star. Similar to the determination of the rotation velocity, all such methods must therefore be able to significantly reduce such interfering side influences mostly by analyzing lines, which are specifically sensitive to temperature [6]. If the temperature should not just be "estimated" but rather accurately "determined," just relatively sophisticated methods remain. Based on high-resolution spectroscopy, such exist for the late spectral classes K–M, by a detailed analysis of especially temperature-sensitive, non-ionized metal lines. A representative impression provides [46], [47], [48]. Here, usually first ratios with the relative line depths $P = I/I_C$ are calculated, at different temperature-sensitive metal absorptions [6]. Subsequently these P-values get calibrated in respect of known T_{eff} values. This way an accuracy of a few kelvin can be achieved. With long term monitoring the detection and even measurement of giant dark sunspots is possible, typically observed at the late spectral class K [48].

10.2.5 The "Balmer Thermometer"

The temperature determination, based on the intensity of the H Balmer lines, is sometimes called the "Balmer thermometer." This method is rather rudimentary, but it provides an interesting experimental field. The H lines are particularly suited for this purpose because the stellar photospheres of most spectral classes consist of >90% hydrogen atoms. Further, exclusively this element can be detected and analyzed within almost the entire temperature sequence (classes O–M). In contrast, ionized calcium Ca II appears only within the spectral classes A–M. The often proposed sodium double line

$D_{1,2}$ is analyzable just at the types ~F–M because Na I becomes ionized at higher temperatures and Na II absorptions appear exclusively in the UV range of the spectrum. In the earlier spectral classes, Na I is therefore always of interstellar origin and hence useless for this purpose.

Figure 10.7 shows the different intensities of the Hβ line, from a cut out of the overview to the spectral sequence in [1]. From all Balmer lines, this absorption can be observed within the largest wavelength range. At this low resolution, it remains analyzable, even in the long wavelength region, to

about class K5. The maximum intensity is reached at the spectral class ~A1. The quantum-mechanical reasons for this effect are discussed in Section 2.4.

Figure 10.8 displays the Hβ equivalent widths (EW) of 24 stars in the *Spectral Atlas* [1], plotted against the according effective temperatures $T_{eff} = f(EW)$. Due to the bell-shaped curve an ambiguity arises, which requires a clarification by additional spectral information. A prominent feature, indicating the profile segment on the long wavelength side of 10,000 K is, for example, the Fraunhofer K line (Ca II) at 3934 Å. The average temperature range is here almost exclusively represented by main sequence stars, the peripheral sections also by giants of the luminosity class I–III. Despite relatively few data points and a low resolution (DADOS 200 L mm^{-1}), the shape of the curve can clearly be recognized.

10.3 Spectroscopic Distance Measurement

10.3.1 Options for Spectroscopic Distance Measurement

Distances can spectroscopically be determined either with the spectroscopic parallax or in the extragalactic range with the help of the Doppler-related redshift, combined with Hubble's law (Section 11.1.10) or by cosmological models. Also, in professional astronomy the empirical Tully–Fisher relationship is applied, which allows estimating the intrinsic luminosity of distant galaxies, based on their angular velocity, measured, for example, by the broadening of emission lines. Within the solar system these spectroscopic methods are supplemented by radar and laser reflectance measurements, in the closer solar neighborhood by the trigonometric parallax and in the Milky Way and extragalactic area by the photometric parallax. The latter

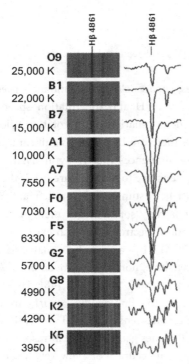

Figure 10.7 The Balmer thermometer with intensities of the Hβ line

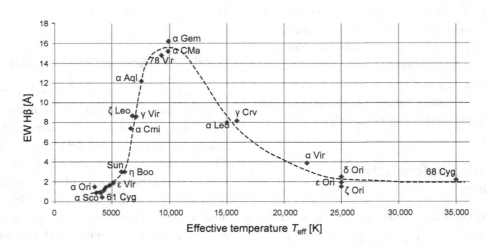

Figure 10.8 Hβ equivalent widths (EW) of 24 stars in the *Spectral Atlas* plotted against the corresponding effective temperature, T_{eff}

is based on the difference between the estimated, absolute visual brightness M_v of an object and its measured apparent magnitude m_v, enabling the estimation of the distance. The absolute magnitude of a stellar object is generally defined for a distance of 10 parsecs [pc] or 32.6 ly.

10.3.2 Term and Principle of Spectroscopic Parallax

For stars, the spectroscopic parallax supplements the photometric by a determination of the spectral class and its estimated absolute visual magnitude M_v. Therefore the term "parallax" is here a misnomer. However, it is appropriate for the trigonometric parallax, which is based on the apparent shift of an observed celestial body, relative to the sky background and caused by the Earth's orbit around the Sun.

A special case of the spectroscopic parallax, applied with considerable success in professional astronomy, is to look for very remote "stellar twins" with nearly identical spectra like those recorded of rather nearby stars with a well-known distance.

10.3.3 Spectral Class and Absolute Magnitude

In Appendix B the average absolute visual magnitudes (M_v) for the spectral classes of the main sequence stars are listed [7]. Important to note for individual stars significant deviations may occur. For giants (III) and super giants (I), this table is supplemented with some values of known stars from the literature in order to give an impression of their considerable scattering range. Therefore within the luminosity classes of the giants, no usable empirical relationship to the spectral classes can be applied. Further, super giants of early spectral classes are very often spectroscopic binaries. These facts clearly demonstrate the limitations of this measuring method. At least for amateurs, the rough distance estimation, applying the spectroscopic parallax, remains mainly restricted to main sequence stars.

10.3.4 Wilson–Bappu Effect

A possible way to also use giants from the spectral classes G–M for the spectroscopic distance estimation is to apply the Wilson–Bappu effect. In 1957, Wilson and Bappu discovered a remarkable correlation between the measured width of the small emission in the core of the Ca II K line ($\lambda 3933.66$) and the absolute visual magnitude of giants. For amateurs with high resolution spectrographs ($R \sim 20{,}000$) this is an

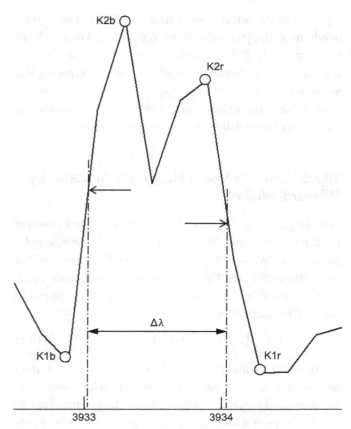

Figure 10.9 Double peak emission core in the K-line of Arcturus (SQUES echelle spectrograph)

interesting field for experiments. Figure 10.9 shows a strong zoom on the emission core of Arcturus, fully displayed in the *Spectral Atlas* [1]. It shows how the width $\Delta\lambda$ is measured, for instance on each side of the line at half intensity between K1 and K2. Here follows, in a long sequence since the 1960s, the most recent calibration [55] of this empirical law:

$$M_v = 33.2 - 18 \log W_0 \qquad \{10.1\}$$

where $W_0 = (\Delta\lambda/\lambda_0)c$. The Ca II K line is preferred for this analysis because it has been revealed that in high resolution spectroscopy the Ca II H line is often contaminated by other adjacent lines. The according spectra have been recorded in La Silla, with the Coudé Echelle Spectrometer and a resolving power of $R \sim 60{,}000$. The analysis was carried out at Gaussian fitted peaks, without consideration of instrumental broadening. The root mean square error of the fitting is indicated with 0.6 mag.

The following example with Arcturus shows very roughly the procedure involved. The measured value is $\Delta\lambda \approx 1.0$ Å. Due to the significantly lower resolution of $R \sim 20{,}000$ it was corrected from the instrumental broadening according to

Equation {9.10}, which results in $\Delta\lambda \sim 0.98$ Å. This corresponds to a Doppler velocity of $W_0 \approx 74.8$ km s^{-1}. With Equation {10.1} it finally yields $M_v \approx -0.53$. The accepted value is -0.3. This formula is highly sensitive in respect of W_0. So for a better accuracy an average of several measurements would be required, in addition to a higher resolved profile and possibly an interpolation of the polygonal intensity line.

10.3.5 Absolute Visual Magnitude Indicator by Millward–Walker

This simple empirical method is a further fast and easy way for the spectroscopic distance estimation of giants, supplementing the Wilson–Bappu based method for giants of the early classes B0–A5. The absolute visual magnitude M_v is here estimated by Equation {10.2}, based on an empirical $M_v = f(H\gamma)$ calibration [56]:

$$M_v = -9.456 + (2 - 0.0942\,s)\,W(H\gamma) \qquad \{10.2\}$$

where $W(H\gamma)$ is the measured EW of the Hγ line in Å and s represents a "spectral index." The value of s refers here simply to the spectral subclass. For example, for the spectral types B0 to B9, the spectral index s goes analogously from 0 to 9, and for spectral classes A0 to 5 it accordingly gives s-values from 10 to 15.

After measuring the EW of the Hγ line, M_v is calculated either with Equation {10.2} or by analysis of its graphical representation. The probable error for a single super giant spectrum is estimated by the authors of the study to just 0.18 mag. The spectra have been recorded ~1984 by a 1872F/30 Reticon diode array on the Cassegrain Spectrograph of the

Table 10.2 Distances to nearby stars

Star	Spectral Class	m_v	M_v	r
Sirius, α CMa	A1Vm	−1.46	1.43	2.64 pc (8.6 ly)
Denebola, β Leo	A3V	2.14	1.93	11.0 pc (36 ly)

DAO 1.88m telescope at a rather low dispersion of 60 Å mm^{-1} corresponding to 0.9 Å diode^{-1}.

10.3.6 Distance Modulus and Estimation of the Distance

Simplified, without any consideration of interstellar extinction, the distance modulus is defined by the difference between the apparent m_v and absolute magnitude M_v, expressed in the generally used logarithmic system of the photometric brightness levels [mag]:

Distance modulus $\mu = m_v - M_v$ [mag]   {10.3}

Simplistically assuming no interstellar extinction, the relationship between the distance r [pc] and the distance modulus $m_v - M_v$ can be expressed as:

$$(m_v - M_v) = 5\log r - 5 \quad [\text{mag}] \qquad \{10.4\}$$

By logarithmic transforming the distance r can be explicitly expressed:

$$r = 10^{0.2(m_v - M_v + 5)} \quad [\text{pc}] \qquad \{10.5\}$$

Examples of distances from Earth for nearby main sequence stars (with published values) can be found in Table 10.2.

11

Expansion and Contraction

11.1 Radial Velocity and Expansion of the Spacetime Lattice

11.1.1 The Radial Velocity

In astrophysics the radial velocity means the motion component of gases, solid substances or celestial bodies, measured on the axis of our line of sight. In the following example (Figure 11.1), observer B measures the radial velocity components v_r of a very closely passing star S, which moves with the velocity v. If the vector v_r is directed towards the observer B, the observed wavelength appears to be compressed and the spectrum appears as blueshifted. In the opposite case it appears to be stretched and the spectrum appears as redshifted. Perpendicular to the line of sight, the apparent transverse velocity v_t of the star can be observed.

11.1.2 The Classical Doppler Effect

Astrophysics uses the classical Doppler principle for the determination of low radial velocities ($\lesssim 500$ km s^{-1}). This law is named after the Austrian physicist Christian Doppler (1803–1853). It applies not only to sound, but also to electromagnetic waves, propagating even in the interstellar vacuum. The classical model explanation of the effect is the changing pitch of a siren heard by an observer as an emergency vehicle passes. The Doppler principle is highly important for astrospectroscopy, so the appropriate equations are derived here and considered in more detail.

Figure 11.2 shows three basic cases, each with one complete oscillation period of a light signal, emitted by a spaceship S with the rest wavelength λ_0 and within the duration T. However, due to the different radial velocities v_r, observer

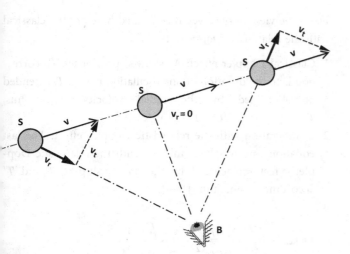

Figure 11.1 Redshift and blueshift due to the radial velocity, v_r

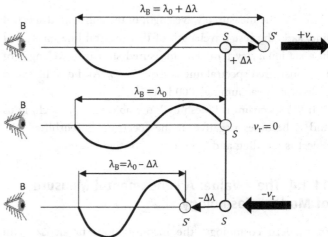

Figure 11.2 Three basic cases of the classical Doppler principle

B measures for all cases different wavelengths λ_B with a correspondingly different duration T_B.

1. In the upper profile the spaceship recedes, during the emission of one entire oscillation period, by the small amount $+\Delta\lambda$ from S to S'. Therefore, the end of the emitted oscillation period compensates this additional way to the observer B with a correspondingly extended duration T_B. Thus B measures a redshift – an extension of the rest wavelength λ_0 by $+\Delta\lambda$. In this case, the sign of the radial velocity v_r is positive (+):

$$\lambda_B = \lambda_0 + \Delta\lambda = cT + v_r T$$
$$T_B = T + (v_r T)/c.$$

2. For the middle profile, observer B measures the originally transmitted oscillation period with the so-called rest wavelength λ_0, because the radial velocity of the spaceship is here $v_r = 0$ and $\lambda_B = \lambda_0$, $T_B = T$.

3. In the lower profile the spaceship approaches by the small amount $-\Delta\lambda$ from S to S' during the emission of one entire oscillation period. Therefore the end of the emitted oscillation period compensates this shortened way to the observer B with a correspondingly reduced duration T_B. Thus B measures a blueshift – a reduction of the rest wavelength λ_0 by $-\Delta\lambda$. In this case, the sign of the radial velocity v_r is negative (–):

$$\lambda_B = \lambda_0 - \Delta\lambda \quad T_B = T - (v_r T)/c.$$

11.1.3 The Spectroscopic Doppler Equation

By the relations: $v_r = \Delta\lambda/T$ and $T = \lambda_0/c$ we obtain the important spectroscopic Doppler equation:

$$v_r = \frac{\Delta\lambda}{\lambda_0}c \quad \text{or} \quad \Delta\lambda = \frac{v_r \lambda_0}{c}. \qquad \{11.1\}$$

Here, λ is the measured wavelength of the considered spectral line, λ_0 is the rest wavelength of this spectral line, measured in a stationary system. The measured shift in wavelength of the considered spectral line is $\Delta\lambda = \lambda - \lambda_0$. And c is the speed of light in a vacuum 300,000 km s^{-1}.

If the spectrum is blueshifted, the object is approaching us and v_r becomes negative. If the spectrum is redshifted, the object is receding and v_r becomes positive.

11.1.4 The z-Value: A Fundamental Measure of Modern Cosmology

In modern cosmology, the inconspicuous break $\Delta\lambda/\lambda_0$ in Equation {11.1} has become the most important measurement variable in the extragalactic space. Now, due to the constant speed of light, this so-called z–value – applied at such extremely distant objects – became both a measure for the distance and for the past. In a calibrated spectrum it can easily be determined by the shift of a spectral line with the measured wavelength λ and their known rest wavelength λ_0. And $\Delta\lambda$ is always proportional to λ_0, whereas z remains constant, i.e. independent of the considered wavelength:

$$z = \frac{\Delta\lambda}{\lambda_0} \quad \text{or} \quad z = \frac{\lambda}{\lambda_0} - 1 \qquad \{11.2\}$$

The z-values also remain completely independent of the debated cosmological model parameters and thus are reliably comparable. According to Equation {11.1}, the radial velocity of the classical Doppler effect results to:

$$v_r = cz \qquad \{11.3\}$$

11.1.5 The Relativistic Doppler Equation

For the determination of higher radial velocities ($\gtrsim 500$ km s^{-1}) such as occur at stellar winds of Wolf–Rayet (WR) stars, nova outbursts or supernova explosions, a modified Doppler equation should be applied, taking in to account the effects of the special theory of relativity (STR). It is based on the different system times T for the observer B and T' for a very fast moving spaceship. According to STR for the relationship of T and T' applies:

$$T = \frac{T'}{\sqrt{1 - (v_r/c)^2}} \qquad \{11.4\}$$

From the view of observer B combined here are the classical with the relativistic Doppler effect:

1. Classical Doppler effect: As showed previously, T_B corresponds to the duration of one oscillation period T, extended or shortened by the radial velocity $\pm v_r$. Thus, $T_B = T + (Tv_r)/c = T(1 + v_r/c)$.

2. Combination with the relativistic Doppler effect: The last equation can now be combined with the relativistic Doppler effect replacing T by the relationship of T and T' according to Equation {11.4}:

$$T_{B\ Rel} = \frac{T'(1 + v_r/c)}{\sqrt{1 - (v_r/c)^2}} = T'\sqrt{\frac{c + v_r}{c - v_r}}$$

After rearranging this equation to isolate v_r and further transforming, finally results in the well-known relativistic Doppler equation:

$$v_r = c \cdot \frac{(z+1)^2 - 1}{(z+1)^2 + 1} \qquad \{11.5\}$$

This applies strictly limited to kinematically induced radial velocities. An extended application to the apparent "cosmological escape velocity," due to the expansion of the "spacetime lattice" is rejected today by experts (see Section 11.1.10).

11.1.6 Measurement of the Doppler Shift and Determination of Radial Velocity

Preliminary remarks: The necessary procedures, theoretical background and hardware for high precision measurements, as required for the search of exoplanets, are beyond the scope of this book. The following is addressed to the average skilled and equipped amateur.

First of all, any steps outlined in Section 8.1.6, to optimize the calibration quality, must be followed. Further, the calibrated spectral lines must be measured as accurately as possible, according to Section 9.1.1. The Doppler shift $\Delta\lambda$ is calculated as difference between the measured wavelength λ of a spectral line and their well-known rest wavelength λ_0:

$$\Delta\lambda = \lambda - \lambda_0 \qquad \{11.6\}$$

In the long wave region of the spectrum and with high-resolution spectrographs, a very accurate calibration can also be achieved by use of the telluric H_2O lines. However, for high accuracy requirements a determination of the systematic measurement error of the whole instrumental setup is inevitable. One possible option is to measure the deviation of a recorded calibration lamp spectrum compared with the unshifted wavelengths of the atmospheric H_2O lines, for example the daylight spectrum [1].

Important is the selection of appropriate spectral lines. The measured Doppler shift is proportional to the according rest wavelength λ_0 (Section 11.1.4). Slim and unblended metal lines, appearing in the long wave range, should therefore be preferred.

With the measured Doppler shift the radial velocity can be determined by Equations {11.1} and {11.5}. For minor accuracy requirements, this measurement can directly be compared with the known radial velocity of a specific standard star that must be located and recorded in the immediate vicinity of the examined object. Such values can, for example, be found in CDS database [25] or in [38]. This allows the compensation of the systematic measurement error and further to avoid a

heliocentric correction which would be required for absolute measurements.

11.1.7 Radial Velocities of Nearby Stars

Due to the common orbit around the galactic center, the radial velocities of stars in the vicinity of the solar system reach mostly just one or two digit values in km s^{-1}. Examples are Aldebaran +54 km s^{-1}, Sirius –8.6 km s^{-1}, Betelgeuse +21 km s^{-1}, Capella +22 km s^{-1} [25]. The corresponding shifts of $\Delta\lambda$ are therefore very low, usually just a fraction of 1 Å. For $\Delta\lambda$ = 1 Å and based on the Hα line (λ6563) Equation {11.1} results in $v_r \approx 46$ km s^{-1}. Thus, such measurements inevitably require highly resolved spectra.

11.1.8 Relative Doppler Shift within a Spectral Profile

Textbook examples for this effect are the P Cygni profiles (Section 11.2.1). For determination of expansion or contraction velocities neither absolutely wavelength calibrated spectra nor any heliocentric corrections are required. A relative measurement of the wavelength difference is sufficient here.

11.1.9 Radial Velocity of Galaxies

In the area between the galaxies, which means outside of strongly gravitational acting systems, the cosmologically induced expansion of the spacetime lattice gets more and more dominant and the kinematic peculiar motion of the galaxies increasingly negligible. This effect must always be taken into account by the interpretation of extragalactic spectra. In the relative "near vicinity" up to ~300 Mly – which also includes Messier's galaxies – it still needs to distinguish between the components of the kinematic Doppler effect, due to the peculiar motion, and the relativistic-cosmological redshift, caused by the expansion of the spacetime lattice. This way, 6 of the 38 Messier galaxies tend to move – against the "cosmological trend" – for instance, with blue-shifted spectra towards our Milky Way! These include M31 (Andromeda) with about –300 km s^{-1}, and M33 (Triangulum) with some –179 km s^{-1} (see Table 11.1).

11.1.10 The Expansion of the Spacetime Lattice

Georges Lemaître (1894–1966) and Edwin Hubble (1889–1953), are considered to be the discoverers of the expanding Universe. Independent of each other they postulated, based on a statistically rather small basis of just 18 galaxies, in relation to photometrically determined distances D, an approximately proportional increase of the redshift z, and the associated radial velocity v_r.

$$v_r = cz = H_{(0)}D \qquad \{11.7\}$$

Where $H_{(0)} \approx 74.3 \pm 2.1$ km s^{-1} Mpc^{-1}, D = Distance [Mpc]. The proportionality factor, the "Hubble constant, $H_{(0)}$" designates the current value of the Hubble parameter $H_{(t)}$ which changes over time. This value is constantly being debated, refined and was determined also by the so-called "$H_{(0)}$ Key Project" with the Hubble Space Telescope. A further attempt was made 2012 with the Spitzer Space Telescope in the mid-infrared, yielding the value applied in Equation {11.7}. In contrast to Lemaître, and somewhat later also Hubble, some researchers believed at that time the Doppler effect was a result of a purely kinematic expansion. So, even today, the term "escape velocity" is still in use. Today, however, it seems clear that, from a distance, longer than say 100 Mpc, the cosmological spacetime expansion is dominating and the phenomenon of redshift has here nothing more to do with the Doppler effect! Due to the expansion and curvature of space, in this extreme distance range, the classical notion of distance, measured in light years (ly) or parsec (pc), gets increasingly problematic and at about >400 Mpc or $z > 0.1$ the Hubble law (Equation {11.7}), should no longer be applied proportionally, for instance, without cosmological model parameters. Therefore the z-value is well established here, which can directly be measured in the spectrum and remains independent of the debated cosmological models (see Section 11.1.12). Therefore, the classical approach to determine the distance by the linear Hubble law remains restricted on the relative "near vicinity" and is still applicable, for example in the realm of the Messier galaxies.

11.1.11 The Apparent Dilemma at $z > 1$

The observable range reaches today up to $z > 10$. However, already for values $z > 1$, already applied in Equation {11.7} ($v_r = cz$), at least an apparent dilemma shows up, because the radial velocity v_r, respectively the expansion of spacetime, seems to surpass the speed of light c. As a possible way out, the application of the relativistic Doppler equation {11.5}, based on the special theory of relativity (STR), is no more accepted today, in contrast to the still debated, different cosmological models – most of them based on Einstein's general theory of relativity (GTR).

11.1.12 The z-Value: Considered as a Measure for the Past

Preliminary remark: Most of the following definitions are according to NASA Extragalactic Database (NED) [26]. The estimated values for the parameters are based on the NASA Wilkinson Microwave Anisotropy Probe Mission (WMAP) launched in 2001.

An in depth consideration of cosmological models is clearly beyond the scope of this book. However, today amateurs also have to interpret large z-values. Therefore the nonlinear relationship to the "look back time" shall roughly be displayed here. It is defined as "The time in the past at which the light we now receive from a distant object was emitted." This relation depends strongly on the assumed cosmological model (Figure 11.3) and can be determined by various online "cosmology calculators," for example provided by NED. In most cases just the currently valid Hubble parameter $H_{(0)}$ and the density parameter Ω are required. In cosmology, Ω is defined as the "Ratio of the average density of mass (ρ) in the Universe to the critical mass density $\rho_{critical}$, the latter being the density of mass needed to eventually halt the outward expansion of the Universe."

So the critical density is defined as $\Omega = \rho/\rho_{critical} = 1$, whereas $\rho_{critical} \approx 10^{-29}$ g m^{-3} corresponding to just ~5 hydrogen atoms per m^3.

$\Omega > 1$: "Closed Universe" with enough matter to reverse the expansion and finally causing a recollapse. Gravity is here strong enough to curve the space back on itself, forming a finite volume with no boundary and therefore non-Euclidian.

$\Omega < 1$: "Open Universe" which does not contain enough matter to halt its expansion. The spacetime geometry is hyperbolic, or "open" and therefore non-Euclidian.

$\Omega = 1$: "Flat Universe" the kinetic energy of the expansion is exactly balanced by the potential gravitational energy of the matter. Thus, after an infinite amount of time the Universe will stop expanding. It has zero curvature to the spacetime continuum and so it is Euclidian.

Today generally a flat Universe with $\Omega \approx 1$ is assumed but is still debated. The total density parameter Ω is mainly determined by the sum of two sub-parameters, wherein the mass density Ω_{rel} of the relativistic particles, made up of electromagnetic energy and neutrinos, can be neglected for our purposes:

$$\Omega = \Omega_m + \Omega_\Lambda \qquad \{11.8\}$$

Here, Ω_m is the mass density including the ordinary (baryonic) mass plus dark matter. This value is estimated today as $\Omega_m \approx 0.28$. And Ω_Λ is the source of anti-gravity, attributed to the dark energy, driving an acceleration of the expansion. This value is estimated today at $\Omega_\Lambda \approx 0.72$.

The required variables for the NED "Cosmology Calculator II," by Nick Gnedin, University of Colorado, are just the z-value of the currently valid Hubble parameter $H_{(0)}$ and the density parameter for baryonic and dark matter Ω_m. Figure 11.4 shows the "look back time" plotted over the z-values for a flat Universe and the default values for $H_{(0)}$ and Ω_m provided by the tool.

11.1.13 Messier Galaxies: Radial Velocity and Cosmological Spacetime Expansion

Table 11.1 shows, sorted by increasing distance D, the measured heliocentric radial velocities v_r and the according z-values for 38 Messier galaxies, after the NED [26]. The cosmological related spacetime expansion v_{exp} is calculated according to Equation {11.7} and $H_{(0)} = 74.3$ with distance values obtained from the Centre de Donnes astronomiques de Strasbourg (CDS) [25].

These figures clearly show, that in this "immediate neighborhood," the kinematic peculiar motions of the galaxies are still dominating. Nevertheless, at distances about more than 50 Mly, the trend becomes already evident that the measured radial velocities v_r with increasing distance tend more and more to the theoretical/cosmological velocity of the spacetime expansion v_{exp}. Anyhow at more than 50 Mly distance, two galaxies (M86 and M98) can still be found with relatively strong negative values (see Table 11.1, gray shaded rows). Six of 38, or about 16%, of these Messier galaxies show this behavior. The most distant galaxy is here M109 which is about 81 Mly.

Relatively considered in the cosmic scale, these very modest amounts show impressively how extremely small this area really is.

11.1.14 The Redshift of Quasar 3C273

The very low z-values listed in Table 11.1 clearly show that Messier's world of galaxies just belongs to our "backyard" of the Universe. As a contrast the apparently brightest quasar 3C273, presented in the *Spectral Atlas* [1], shows with $z = 0.158$ impressively redshifted H emission lines. Thanks to advances, particularly in camera technology, amateur astronomers nowadays also have the pleasure to deal with cosmologically relevant distance ranges. We must therefore be aware of the effects of Einstein's GTR, as well as of the cosmological models, which, however, are still under debate. Quasar 3C273 also demonstrates that the equations presented earlier have even for amateurs a real, practical application.

11.1.15 The Gravitational Redshift or Einstein Shift

Besides the classical Doppler effect and the relativistic space-time expansion, there remains a third option – the gravitational redshift or Einstein shift. According to the GTR, light measurably loses energy as it propagates in a strong gravitational field. The extreme cases here are neutron stars or even black holes. In the spectroscopic amateur practice, however, already the much smaller gravitational field of a white dwarf can massively distort the measurements of radial velocity [1]. The following formula shows the gravitational redshift, predicted by the GRT:

$$\frac{\lambda}{\lambda_0} = \frac{1}{\sqrt{1 - \frac{2GM}{rc^2}}} \qquad \{11.9\}$$

Here, G is the gravitational constant (see Section 12.2.7), M is the mass of the celestial object, r is the radius between the center of the object and the point where the light is emitted and c is the velocity of light.

11.1.16 Age Estimation of the Universe

Assuming a constant expansion rate of the Universe and by rearranging Equation {11.7}, the time span since the big bang can roughly be estimated when the entire matter was concentrated at "one point."

$$\frac{1}{H_0} = \frac{D}{v_{exp}} = t_{H(0)} \qquad \{11.10\}$$

This lapse of time is also called Hubble time $t_{H(0)}$ and defined as the reciprocal Hubble constant. This is also equal to the division of the distance D by the expansion velocity v_{exp}, indicating that we obtain here the estimated age of the Universe. To calculate the "Hubble time" we need first to put the units into the reciprocal term of the Hubble constant. To obtain the age of the Universe in years instead of seconds we have to divide this term by the number of seconds per year ($3.15 \times 10^7 s$). Further, to eliminate the unused units for the distance we have to convert Mpc into km (1 Mpc $= 3.09 \times 10^{19}$ km):

Table 11.1 Messier galaxies, distances, z-values, radial velocities and cosmologic spacetime expansion (positive values = redshifted, negative values = blueshifted)

Messier galaxy	Distance D [Mpc/Mly]	z-value	Radial velocity v_r [km s^{-1}]	Cosmological spacetime expansion v_{exp} [km s^{-1}]
M31 Andromeda	0.79/2.6	−0.0010	−300	+59
M33 Triangulum	0.88/2.9	−0.0006	−179	+65
M81	3.7/12	−0.0001	−34	+275
M82	3.8/12	+0.0007	+203	+282
M94	5.1/17	+0.0010	+308	+379
M64	5.3/17	+0.0014	+408	+394
M101	6.9/22	+0.0008	+241	+513
M102	6.9/22	+0.0008	+241	+513
M83	7.0/23	+0.0017	+513	+520
M106	7.4/24	+0.0015	+448	+550
M51 Whirlpool	8.3/27	+0.0020	+600	+617
M63	8.3/27	+0.0016	+484	+617
M74	9.1/30	+0.0022	+657	+676
M66	10.0/32	+0.0024	+727	+743
M95	10.1/33	+0.0026	+778	+750
M104 Sombrero	10.4/34	+0.0034	+1024	+773
M105	10.4/34	+0.0030	+911	+773
M96	10.8/35	+0.0030	+897	+802
M90	12.3/40	−0.0008	−235	+914
M65	12.6/41	+0.0027	+807	+936
M77	13.5/44	+0.0038	+1137	+1003
M108	14.3/47	+0.0023	+699	+1062
M99	15.4/50	+0.0080	+2407	+1144
M89	15.6/51	+0.0011	+340	+1138
M59	15.6/51	+0.0014	+410	+1138
M100	15.9/52	+0.0052	+1571	+1160
M98	16.0/52	−0.0005	−142	+1181
M49	16.0/52	+0.0033	+997	+1181
M86	16.2/53	−0.0008	−244	+1204
M91	16.2/53	+0.0016	+486	+1204
M60	16.3/53	+0.0037	+1117	+1211
M61	16.5/54	+0.0052	+1566	+1226
M84	16.8/55	+0.0035	+1060	+1248
M87	16.8/55	+0.0044	+1307	+1248
M85	17.0/55	+0.0024	+729	+1263

Table 11.1 (*cont.*)

Messier galaxy	Distance D [Mpc/Mly]	z-value	Radial velocity v_r [km s^{-1}]	Cosmological spacetime expansion $v_{\exp}$ [km s^{-1}]
M88	18.9/62	+0.0076	+2281	+1404
M58	19.6/64	+0.0051	+1517	+1456
M109	24.9/81	+0.0035	+1048	+1850

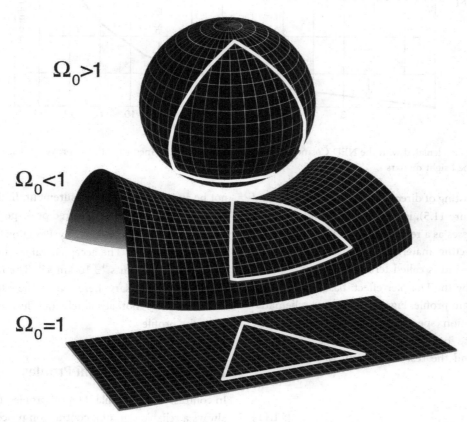

$\Omega_0 > 1$

$\Omega_0 < 1$

$\Omega_0 = 1$

Figure 11.3 Geometry of the Universe depending on the density parameter, Ω. (Credit: NASA/WMAP Science Team)

$$t_{H(0)} = \frac{[s] \cdot [Mpc]}{74\,[km]} = \frac{3.09 \cdot 10^{19}\,[km]}{74\,[km] \cdot 3.15 \cdot 10^7}$$

$$= 1.34 \cdot 10^{10}\,\text{yr or } 13.4\text{ bn years}$$

Due to the assumed constant expansion rate the Hubble time $t_{H(0)}$ is slightly different to the accepted age of the Universe of 13.7 billion years.

11.2 Measurement of Expansion and Contraction

At certain stages of evolution some stars eject matter. The strength of the ejection varies as does the speed, which ranges from a relatively slow 20–50 km s^{-1} typical for planetary nebulae, up to several 1000 km s^{-1} for novae, supernovae and WR stars. Such processes manifest themselves by different spectral symptoms, dependent mainly on the density of the ejected material. Under other circumstances, in some objects that are forming accretion disks or in the stellar atmospheres of some giants (e.g. Rigel), sometimes contraction processes can be observed.

11.2.1 P Cygni Profiles

The generation of P Cygni profiles is explained in the *Spectral Atlas* [1], at the example of the LVB star P Cygni. This

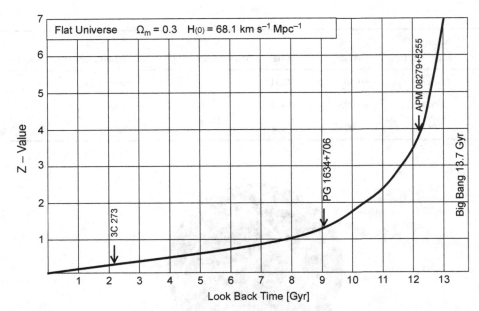

Figure 11.4 Plotted graph calculated with the NED Cosmologic Calculator II (by Nick Gnedin) with the provided default values and supplemented with some bright quasars

spectral feature, consisting of directly connected emission and absorption parts (Figure 11.5), is very common and occurs in nearly all spectral classes as a reliable sign for relatively dense, expanding or contracting material. The maximum velocity reached by a stellar wind is called terminal velocity v_∞ [45]. It can be estimated by the Doppler effect, based on a measurement in the P Cygni profile. Figure 11.5 shows schematically for high resolution spectra the measurement of the wavelength difference between the "blue edge" λ_∞ of the absorption trough and the rest wavelength λ_0 of the entire spectral line:

$$v_\infty = \frac{\lambda_\infty - \lambda_0}{\lambda_0}\, c \qquad \{11.11\}$$

However, λ_∞ is easy to determine just at saturated profiles which are mostly found in the UV range and unreachable for amateurs. In the unsaturated case the "blue absorption edge" is located, somewhat diffuse, on the upper part of the blue wing. Alternatively in lower resolved spectra, like displayed in Figure 11.6, for a rough estimation of v_∞, the relatively easily determinable peak-to-peak difference between absorption and emission may be used [13].

This is demonstrated here at the expansion velocity of the stellar envelope of P Cygni. As a rough approximation the offset $\Delta\lambda$ [Å] is measured here between the peaks of the emission and the blueshifted absorption part of the P Cygni profiles [13]. However, the stunted absorption trough would

not be appropriate for a measurement. In the example for the Hα line of P Cygni, the measured peak–peak difference yields: $\Delta\lambda = \lambda_{\text{Blue}} - \lambda_{\text{Red}} \approx -4.5$ Å. By Equation {11.1} it results in $v_\infty \approx -206$ km s^{-1}. The accepted values are within the range of -185 to -205 km s^{-1} ± 10 km s^{-1}. The heliocentric correction is not necessary here, since the Doppler shift $\Delta\lambda$ is measured here not absolutely but just relatively, obtained from the profile.

11.2.2 Inverse P Cygni Profiles

In contrast to the normal P Cygni profiles, the inverse ones are always a reliable sign of a contraction process. The absorption kink of this feature is shifted here to the red side of the emission line. A textbook example is a protostar T Tauri which is formed by accretion from a circumstellar gas and dust disk. The forbidden [O I] and [S II] lines show here strikingly inverse P Cygni profiles, indicating large-scale contractions within the accretion disk. The Doppler analysis shows here contraction velocities of some 600 km s^{-1}. Figure 11.7 shows a detail of the T Tauri spectrum [1]. A similar effect is also exhibited in the spectrum of Quasar 3C273 [1].

11.2.3 Broadening of the Emission Lines

P Cygni profiles are also observable in spectra of relatively dense and rapidly expanding explosion shells of novae and

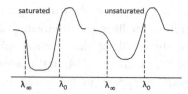

Figure 11.5 Measurement of the absorption trough in high resolution spectra

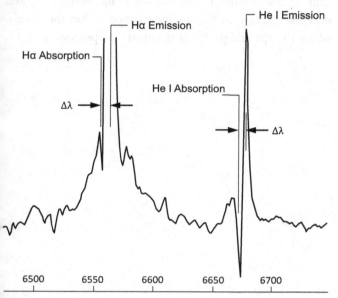

Figure 11.6 Measurement of the peak–peak difference in lower resolved spectra

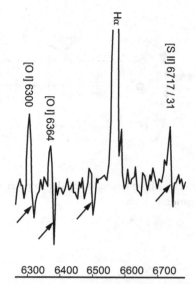

Figure 11.7 T Tauri inverse P Cygni profiles

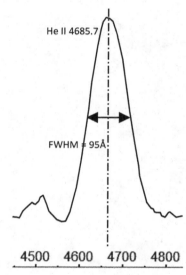

Figure 11.8 Broadened He II emission of WR142

---

supernovae. More often such extreme events show just a strong broadening of the emission lines. This applies also to Wolf Rayet stars and quasars. The expansion or terminal velocity v_∞ can be estimated here with the conventional Doppler equation, if $\Delta\lambda$ is simply replaced by $\text{FWHM}_{\text{Emission}}$:

$$v_\infty \approx \frac{\text{FWHM}_{\text{Emission}}}{\lambda_0} c \qquad \{11.12\}$$

Figure 11.8 shows the near perfectly Gaussian-shaped He II emission line of WR142 at λ4685 [1]. When Equation {11.12} is applied the expansion velocity yields ~6000 km s^{-1}! Considering some blends, contributing to the line shape and the instrumental broadening, for this extreme Wolf Rayet star, classified as WO2, v_∞ may be estimated to ≳5000 km s^{-1}. Analyzing spectra of nova outbursts [1], in Equation {11.12} instead of the FWHM value, mostly

HWZI (Half Width at Zero Intensity, Section 9.1.14) is applied.

11.2.4 Splitting of the Emission Lines

Split emission lines can be observed in spectra of expanding, transparent shells, for example of old supernova remnants, novae and planetary nebulae. Such examples are presented in [1].

Figure 11.9 explains schematically the split up of the emission lines due to the Doppler effect. The hemisphere

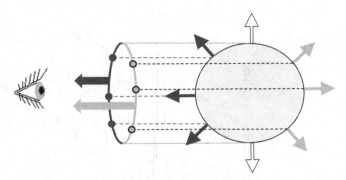

Figure 11.9 Velocity ellipse generated by the Doppler effect in transparent expanding shells

of a more or less transparent shell moving towards Earth causes a blueshift of the lines and a retreating shell causes a redshift. If the slit length of the spectrograph covers the full diameter of the expanding SNR, in the recorded spectral stripe the emissions appear to be spread to a so-called "velocity ellipse." This effect can be seen, significantly less perfect, at the [O III] lines generated by the supernova remnant M1, see [1] and Figure 7.2. The amount of the split $\Delta\lambda$ corresponds to the twofold radial velocity $2v_r$ and depends strongly on the specific location within the nebula, where the spectral profile is obtained and processed.

12 Rotation and Orbital Elements

12.1 Measurement of Rotational Velocity

12.1.1 Terms and Definitions

Due to the Doppler effect a spectroscopic observation of a rotating celestial body in the solar system reveals different apparent radial velocities, if measured at the eastern and western limb. The measured difference Δv_r allows the determination of the rotational induced surface velocity. The focus here is on the spectroscopically direct measurable share, the so-called "$v \sin i$" value, which is projected into the line of sight to Earth. Figure 12.1 shows the relation of the velocity components in function of the inclination angle i between the rotation axis and the line of sight to Earth.

If considered as an equation, the term "$v \sin i$" allows the calculation of this projected velocity share with given equatorial velocity v_e and inclination angle i. Only in the special case if $i = 90°$ and $\sin i = 1$, we can measure the equatorial velocity v_e. If $i = 0°$ we look directly on a pole of the celestial body and thus the projected rotational velocity yields $\sin i = 0°$.

12.1.2 The Rotational Velocity of Apparent 3D Objects

Observed by telescopes large planets, the Sun and even galaxies appear as 3D objects. A quite easy determination of the rotational velocity is enabled here by the determination of the difference in velocity between the eastern and western limb of the rotating object. If recorded at the same time to this relative measurement a heliocentric correction is unnecessary. However, two different sub-cases must be distinguished here.

12.1.2.1 Light-Reflective Objects of the Solar System
Analogously to the radar principle, for light-reflecting objects of our solar system – for example planets or moons – the Doppler effect, observed from Earth, acts twice (see Figure 12.2). A virtual observer, located at the western limb

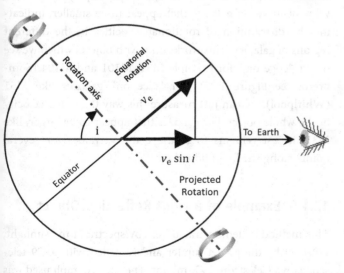

Figure 12.1 The projected rotation (surface) velocity: $v \sin i = v_e \sin i$

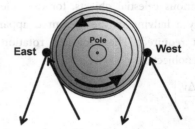

Figure 12.2 The Doppler effect acting twice for reflected light from planets

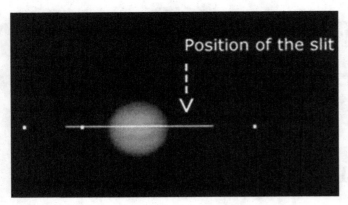

Figure 12.3 Jupiter with position of the slit (M. Trypsteen)

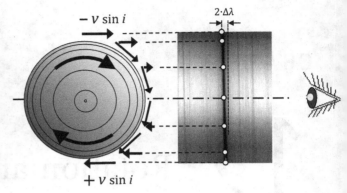

Figure 12.4 Inclination of planetary spectral lines due to the rotation velocity

of the planet, sees the incident light of the Sun as already redshifted by an amount corresponding to the rotationally induced radial velocity of this surface point. This observer also notes that the light is reflected unchanged towards Earth with the same redshift $\Delta\lambda$. Finally, an observer on Earth measures this reflected light as redshifted by an additional amount, which corresponds to the radial velocity of the western limb of the planet, relative to the Earth. When the outer planets are close to the opposition, both of these redshifted amounts are virtually equal, finally resulting in $2 \times \Delta\lambda$. This shift now allows us to calculate an apparent difference in radial velocity Δv_r between the eastern and western limb of the planet. A halving of this value yields the true projected difference in radial velocity. If this is halved again we obtain the required, apparent rotational velocity $v \sin i$. This corresponds finally to one quarter of the originally measured apparent difference in radial velocity Δv_r.

The projected rotation velocity of light-reflecting objects is:

$$v \sin i_{\text{refl}} = \left| \frac{\Delta v_r}{4} \right| \quad \{12.1\}$$

where Δv_r is based here on the effectively measured value of $2 \times \Delta\lambda$.

12.1.2.2 Self-Luminous Celestial Bodies

For self-luminous celestial objects, for example the Sun or galaxies, only a halving of the measured apparent velocity difference is required. The projected rotation velocity of self-luminous objects:

$$v \sin i_{\text{self}} = \left| \frac{\Delta v_r}{2} \right| \quad \{12.2\}$$

where Δv_r is based here on the effectively measured value of $1 \times \Delta\lambda$.

12.1.3 The Rotational Velocity of Large Planets

Spectrographs with longer slits allow the visualization of this Doppler induced effect and further enable even a kind of "graphic" determination of $v \sin i$. If the slit is aligned with the equator of a rotating planet, the absorption lines of the reflected solar light appear slightly slanting. Thereby the atmospheric H_2O lines are not affected and remain vertical. With Jupiter the slit can be roughly aligned with the help of its moons (Figure 12.3) and with Saturn by its ring. The value $2 \times \Delta\lambda$ can be determined either by counting of pixels or by calculation of the velocity difference on the bottom and top of the spectral stripe. Figure 12.4 shows this effect for the case with reflected sunlight.

12.1.4 The Rotational Velocity of Galaxies

The latter method can also be applied to display the famous velocity curves of galaxies that appear to be smaller, indicating the distribution of rotational velocities in the different regions of galaxies. This works only with objects which we see rather "edge on," for example M31, M101 and M104 (Sombrero; see Figure 12.5). For "face on" galaxies like M51 (Whirlpool), we can just measure this way the radial velocity of the whole object. For galaxies that appear to be larger, like M31, it just remains to measure the v_r values for several points along the longitudinal axis.

12.1.5 Example of a Light Reflecting Object

This method is demonstrated here by spectra of the sunlight, reflected by the planet Jupiter and recorded with a C9 telescope on a Celestron AVX mount. The spectrograph used was a Lhires III, equipped with a 2400 L mm^{-1} grating and a

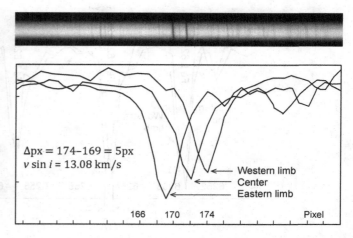

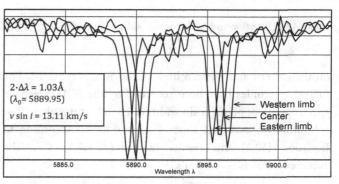

Figure 12.5 M104 with position of the slit. (Background image: NASA/ESA and The Hubble Heritage Team STScI/AURA)

cooled full spectrum modified Canon 600D DSLR camera. The Doppler shift of the well-known sodium doublet D1, D2 at $\lambda\lambda 5895.92, 5889.95$ was analyzed to calculate the projected rotational velocity according Equation {12.1} with measurements based on wavelength (wavelength method). Alternatively Equation {12.3} can be applied with measurements based on counting pixel values (pixel method).

$$v \sin i_{\mathrm{refl}} = \frac{c}{4} \; \frac{\Delta\mathrm{px}\; \mathrm{Disp}}{\lambda_0} \qquad \{12.3\}$$

where c = speed of light, $\Delta\mathrm{px}$ represents the pixel based Doppler shift, λ_0 the rest wavelength of the investigated spectral line and finally Disp is the corresponding dispersion in [Å/pixel]. Based on the $v \sin i$ and the known period of rotation by Equation {12.4} even the diameter of the planet can be estimated.

$$D = \frac{v \sin i \, P}{\pi} \qquad \{12.4\}$$

where D is the diameter of the planet [km], $v \sin i$ the measured, projected rotational velocity [km s^{-1}], P the period of rotation [s] and $\pi = 3.14159$.

Figure 12.6 shows for Jupiter on the top the recorded spectral stripe with the slanted appearing sodium absorptions. In the middle follows, based on the pixel count, the VSpec analysis to estimate the $v \sin i$ value. On the bottom the same was attempted, based on a wavelength calculation and analyzed with the BASS project. In both cases the eastern and western limb have been measured and also the center of the disk.

In any case it is strongly recommended to make several spectral recordings which can be used to calculate the mean velocity value and corresponding standard deviation, which

Figure 12.6 Doppler shift of the sodium D1, D2 absorptions in spectra recorded at the eastern and western limb of planet Jupiter, processed in VSpec and BASS

reflects the precision of the measurements. As seen from Figure 12.6 and based on several measurements the mean calculated projected rotational velocity of planet Jupiter is 13.10 ± 0.03 km s^{-1}. Correspondingly, Equation {12.4} gives the diameter of the planet as 147,716 km. Taking into account the different margins of error, which influence the accuracy, as the inclination of the planet (sin i), influence of the position of the slit and other possible measurement errors, this result is pretty close to published values of rotational velocity of 12.7 km s^{-1} and diameter of 142,984 km [44].

12.1.6 Example of a Self-Luminous Celestial Body

This method also allows the estimation of the Sun's rotational equatorial speed, recorded of course with an attached energy filter(!). It was first practiced 1871 by Hermann Vogel. However, typical for spectral class G, this velocity is very low (~1.9 km s^{-1}), so a high resolution spectrograph is required. In this case, the focal picture of the Sun on the slit plate would be too large to cover the entire solar equator. In such cases the recording of two separate wavelength-calibrated spectra, each

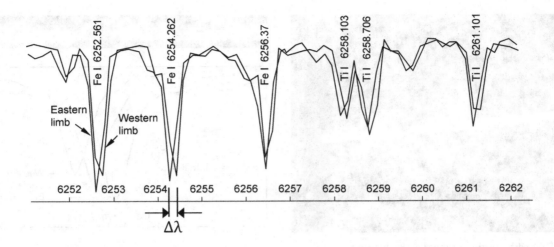

Figure 12.7 Rotation-induced Doppler shift in the spectra from the eastern and western limb of the Sun, SQUES echelle spectrograph

on the eastern and western limb of the Sun, is required. The following profiles have been recorded by the SQUES echelle spectrograph applying a slit width of 20 μm, and Atik 314L+, exhibiting an averaged shift of $\Delta\lambda \approx 0.07$ Å. According to Equation {12.2} this results in $v \sin i \approx 1.7$ km s^{-1}. The accepted value is ~ 1.9 km s^{-1}.

12.1.7 The Rotational Velocity of Stars

Even with large telescopes, stars cannot be seen as 3D objects. They appear as small, so called Airy disks, generated by diffraction effects in the optics. In this case the method, presented earlier, must inevitably fail. Today, professional astronomy applies many sophisticated methods to determine the rotational speed of stars, for example with photometrically detected brightness variations, by means of interferometry or Fourier analysis. For amateurs, we present the rather simple method of measuring the FWHM of rotationally broadened lines.

In 1877 William Abney was the first to propose the determination of $v \sin i$ by analyzing the rotationally induced broadening and flattening of the spectral lines. This phenomenon is induced by the Doppler effect, because the recorded light of the star is emitted by surface points with very different radial velocities. Figure 12.8 demonstrates the principle of how a highly resolved profile of a fine metal line gets flatter and broader with increasing rotational velocity [14].

The numerous measured rotational velocities of ordinary main sequence stars show a remarkable distribution in terms of spectral classes. Figure 12.9 shows a significant decrease in velocity from early to late spectral types. This phenomenon is

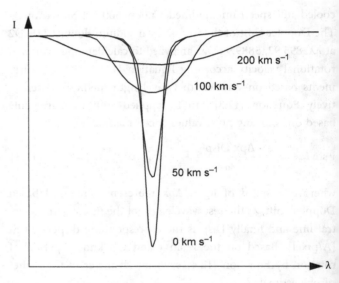

Figure 12.8 Rotationally induced broadening of a fine metal line, schematically after F. Royer (in Rozelot and Neiner [14])

mainly caused by the braking effect of stellar magnetic fields. They get stronger the later the star is classified due to progressively deeper reaching convection zones. For spectral type G and later, $v \sin i$ amounts mostly to just a few km s^{-1}. The entire velocity range extends from 0 to >400 km s^{-1}.

Since, for the spectral classes G and later, the $v \sin i$ values get very low (so-called slow rotators), the measurement here requires spectrographs with high resolution power. Particularly for amateurs the focus of this method is therefore on the early spectral classes A and B which are dominated by the so-called "fast rotators." Various studies have shown that the orientation of the stellar rotation axes is randomly distributed. Since the inclination i of stars is difficult to determine,

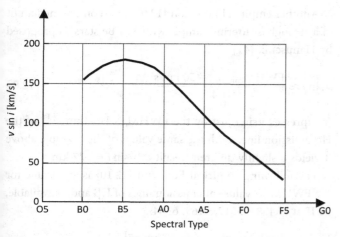

Figure 12.9 Distribution of rotational velocities in function of the spectral class, (schematically after Slettebak [43])

the equatorial velocity v_e is known only in exceptional cases. Therefore the research is here mostly limited to statistical surveys, based on extensive v_r samples. Most amateurs will here be limited, trying to reproduce literature values for early spectral types with high rotational velocities.

12.1.8 Empirical Equations for $v \sin i$ in Function of FWHM

As already shown "rotational broadening" is not the only effect influencing the FWHM of the spectral lines. It was therefore necessary to find methods or particular sensitive lines to isolate the share of rotational broadening. Since the 1920s until the present a considerable number of astrophysical publications have dealt with the calibration of the rotational velocity, relative to FWHM. From 1975 probably the most cited standard work is the "new Slettebak system." However, recent studies have shown that the provided $v \sin i$ values are systematically too low. Therefore more recent approaches are presented here. In such equations, the most requested variable, is $FWHM_{corr}$, adjusted by the instrumental broadening (see Section 9.1.11). Mostly this value is needed in ångstroms, but sometimes expressed also as Doppler velocity [km s^{-1}]. In other cases, the equivalent width EW [Å] is requested.

12.1.9 Calibration Equations by F. Fekel

This method [39], [40] is based on two different calibration curves, each one for the red and the blue region of the spectrum. Two polynomial equations calibrate the "raw

value" X for $v \sin i$ in [km s^{-1}], relating to the measured and adjusted $FWHM_{corr}$ in [Å]. They are presented somewhat transformed here, in order to provide the explicitly requested X-values for the spectral ranges around $\lambda6430$ and $\lambda4500$:

$$X_{6430} \approx \sqrt{7143(FWHM_{corr\ 6430} + 1.08)} - 89.6 \qquad \{12.5\}$$

$$X_{4500} \approx \sqrt{9091(FWHM_{corr\ 4500} + 0.29)} - 58.1 \qquad \{12.6\}$$

An overview of the procedure is presented next.

First, a sample of FWHM values [Å] is measured at weakly to moderately intense metal lines, as proposed below (no H-Balmer absorptions!). Further these values must be cleaned of the instrumental broadening $FWHM_{corr}$ (Section 9.1.11). Then the X-values are calculated by inserting the $FWHM_{corr}$ values in Equation {12.5} or {12.6}, according to the wavelength range.

Further, these X-values must finally be cleaned from the line broadening due to the average macro turbulence velocity v_m in the stellar atmosphere, applying Equation {12.7}. This will yield the requested $v \sin i$ value in [km s^{-1}]:

$$v \sin i = \sqrt{X^2 - v_m^2} \qquad \{12.7\}$$

where the variable v_m depends on the spectral class.

For the B and A class Fekel assumed $v_m = 0$, and for the early F classes $v_m = 5$ km s^{-1}, Sun-like dwarfs $v_m = 3$ km s^{-1}, K-dwarfs $v_m = 2$ km s^{-1}, early G giants $v_m = 5$ km s^{-1}, late G and K giants $v_m = 3$ km s^{-1} and F–K subgiants $v_m = 3 - 5$ km s^{-1}.

12.1.10 Suitable Metal Lines for the FWHM Measurement

Here, the commonly used metal lines for determining the FWHM values are listed. The lines, also proposed by F. Fekel are written in italics, and those used by other authors (e.g. Slettebak) in normal letters. The supplement "(B)" means that the profile shape is blended with a neighboring line and "(S)" means a line deformation due to an electric field by the Stark effect.

– Late F, G and K classes: Lines preferably in the range at $\lambda6430$ Å: e.g. Fe II 6432, Ca I 6455, Fe II 6456, Fe I 6469, Ca I 6471.
– Middle A classes and later: Moderately intense *Fe I, Fe II* and *Ca I lines* in the range at $\lambda6430$. For A3–G0 class: Fe I 4071.8 (B) and 4072.5 (B).
– O, B and early A classes: Lines in the range at $\lambda4500$ Å:

– Middle B to early F classes: Several *Fe II* and *Ti II lines*, as well as *He I 4471* and *Mg II 4481.2*.
– O, early B and Be classes: He I 4026 (S), Si IV 4089, He I 4388, He I 4470/71 (S, B), He II 4200 (S), He II 4542 (S), He II 4686, *Al III, N II*.

Alternatively to F. Fekel, in the context of K giants, A. Moskovitz [42] analyzed exclusively the well isolated Fe I line at λ5434.5 applying Equation {12.5}.

12.1.11 Rotational Velocity of Circumstellar Disks around Be Stars

The Be stars and their spectra are presented in [1]. Many amateurs are involved here in research programs, mainly by spectroscopic monitoring of the Hα emission. Here follow some procedures to estimate the rotational velocity, chiefly by processing of measured EW and FWHM values.

At Be stars the Doppler-broadening of the emission lines is mainly caused by the rotating circumstellar disk of gas. Therefore the corresponding FWHM values are now a measure for the rotational velocity of the ionized disk material. In the time span between the onset of the outburst at δ Scorpii in 2000 and 2011, the FWHM–Doppler velocity of the Hα emission fluctuated between about 100 and 350 km s^{-1} and the EW value from −5 to −25 Å. This indicates highly dynamic processes in the disk formation process. Furthermore the chronological intensity sequence of FWHM and EW values appear strikingly shifted in phase.

12.1.12 Empirical Equations for the Rotational Velocity of the Disk

Several equations have been published, which allow us to roughly estimate the rotation speed of the disk material from Be stars. In most cases $v \sin i$ is calculated with FWHM$_{corr}$ values at the Hα line.

An empirical equation by Dachs *et al.* [92] expresses explicitly the $v \sin i$ value, based on the FWHM$_{corr}$ at the Hα emission line in [km s^{-1}], and the (negative) equivalent width EW [Å]:

$$v \sin i \pm 30 \text{ km s}^{-1} \approx \frac{\text{FWHM}_{\text{corr Hα}}}{2} \left[\frac{\text{EW}_{\text{Hα}}}{-3 \text{ Å}} \right]^{\frac{1}{4}} - 60 \text{ km s}^{-1} \quad \{12.8\}$$

An example here is the measurement of Hα line that yields FWHM$_{corr Hα}$ = 7 Å, corresponding to a Doppler velocity of 319 km s^{-1} and EW = −17 Å. This results in $v \sin i$ = 186 km s^{-1}.

Another empirical Equation {12.9}, based on the median fit of a strongly scattering sample with 115 Be stars, is proposed by Hanuschik [41]:

$$v \sin i \approx \frac{\text{FWHM}_{\text{corr Hα}} - 50 \text{ km s}^{-1}}{1.4} \quad \{12.9\}$$

It expresses $v \sin i$ just by the FWHM$_{corr}$ Hα [km s^{-1}] of the Hα emission line. With the same values of the example above it yields a slightly different result of $v \sin i$ = 177 km s^{-1}.

The following empirical Equation {12.10} is applicable for the FWHM$_{corr}$ values at emission lines of Hβ and, if available, Fe II at λλ4584, 5317, 5169, 6384:

$$v \sin i \approx \frac{\text{FWHM}_{\text{corr Hβ, FeII}} - 30 \text{ km s}^{-1}}{1.2} \quad \{12.10\}$$

12.1.13 Distribution of the Rotational Velocity within the Disk

Assuming that the disk rotation obeys the kinematic laws of Kepler, the highest rotational velocity v occurs on its inner edge, in many cases identical with the star's equator, and decreases towards the outside [13](see Figure 12.10):

$$v = \sqrt{\frac{GM}{R}} \quad \{12.11\}$$

Where G is the gravitational constant, $6.674 \times 10^{-11} \text{ m}^3 \text{ kg}^{-1} \text{ s}^{-2}$, M is the mass of the star [Kg] and R is the distance from the considered disk part to the star's center [m]. The application of this equation is limited to cases with known inclination angle i, allowing the computation of equatorial velocity v_e, or very high $v \sin i$ values with $v \sin i \approx v$.

12.1.14 Analysis of Double Peak Profiles

The emission lines of Be stars often show a double peak. The gap between the two peaks is explained with the Doppler,

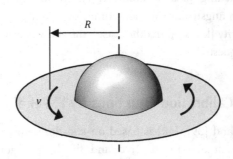

Figure 12.10 Rotation velocity within the disk

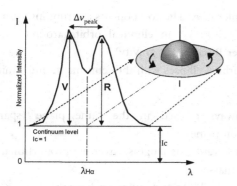

Figure 12.11 Double peak of Hα in relation to the rotating gas disk

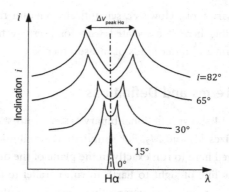

Figure 12.12 Δv_{peak} as a function of the inclination i, schematically after [13]

self-absorption and perspective effects. Figure 12.11 illustrates the double peak in the Hα emission line, in relation to the rotating gas disk. Introduced here are important dimensions, which are used in the literature:

V= Intensity (V/I_C) violet shifted peak

R = Intensity (R/I_C) red shifted peak

I_C = Normalized continuum level = 1

V/R = Peak intensity ratio

Δv_{peak} = Distance between the peaks [km/s^{-1}].

12.1.14.1 Peak Intensity Ratio V/R (Violet/Red)

The V/R ratio is one of the main criteria for the description of the double peak emission in spectra of Be stars. In Be binary systems a pattern of variation seems to occur, which is linked to the orbital period of the system[93]. At δ Scorpii, after the outburst of 2000, the V/R ratio of the He I line (λ6678.15) shows a strong variation of some 15 Å.

- According to Kaler [11] the V/R ratio also reflects the mass distribution within the disk and may show a fairly irregular course.
- In Be binary systems a pattern of variation seems to occur, which is linked to the orbital period of the system.
- According to Hanuschik [41] asymmetries of the emission lines (V/R $\neq$ 1), are related to radial movements and darkening effects.
- If no double peak appears in the emission line the asymmetry in the steepness of the two flanks can be analyzed. And V>R means the violet flank is steeper and vice versa by R>V the red one.

12.1.14.2 Distance between the Peaks Δv_{peak}

Figure 12.12 shows, schematically after [13], the modeled emission lines for different inclination angles i. It seems obvious that the distance Δv_{peak} increases with growing inclination i. At the same time also the $v \sin i$ values are increasing, if for all inclination angles a similar equatorial velocity v_e and a fixed disk radius Rs are assumed. Δv_{peak} is expressed as velocity according to the Doppler principle, analogously to Equation {9.9}.

Also Hanuschik [41] has shown, that a rough correlation exists between the $\Delta v_{\text{peak Hα}}$ [km s^{-1}] and $v \sin i$:

$$\Delta v_{\text{peakHα}} \approx (0.4 \ldots 0.5)\, v \sin i \qquad \{12.12\}$$

According to Miroshnichenko and Hanuschik a decreasing disk radius R is also associated with, an increasing $\Delta v_{\text{peak Hα}}$.

12.1.15 The Outer Disk Radius Rs

The equation by Huang [94] shows the relation between the radius of the disk Rs (Figure 12.13), the stellar radius r, the apparent rotation velocity $v \sin i$ and Δv_{peak}:

$$\frac{\text{Rs}}{r} \approx \left(\frac{2\, v \sin i}{\Delta v_{\text{peak}}} \right)^2 \qquad \{12.13\}$$

The application of this equation is limited to cases with known inclination angle i, allowing the computation of equatorial velocity v_e, or very high $v \sin i$ values i.e. $\Delta\lambda = 0$.

12.2 Estimation of Orbital Elements in Binary Systems

12.2.1 Introduction

Here the rough estimation of orbital elements and stellar masses are presented. It is intended to show what can be achieved with average amateur means and skills, restricted on a spectroscopic assessment, just by a time series of recorded profiles. An in-depth study would be highly demanding and require, for example, profound skills in celestial mechanics and the necessary tools. However, scientifically relevant results are only possible here if associated with long-term astrometric, photometric and other kinds of measurements.

12.2.2 Terms and Definitions

Figure 12.14 shows a fictional binary system with stars of the unequal sizes M_1 and M_2. For simplicity their elliptical orbits are assumed here to run exactly in the plane of the drawing as well as the line of sight to Earth to run parallel to the semi-minor axes b. For this perspective special case the orbital velocity v_M at the apastron (farthest orbital point) and periastron (closest orbital point) corresponds also to the observed radial velocity v_r. In the literature these recorded maximum values (amplitudes) are referred to as K.

12.2.3 Some Basics of Celestial Mechanics

We present here some basics of celestial mechanics, just selected for the understanding of the following methods for the spectroscopic analysis. In contrast to the complex behavior of multiple systems, the motion of binary stars follows the

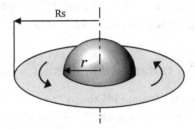

Figure 12.13 Outer disk radius Rs

three Kepler laws. The components are running with variable track velocities v_M in elliptical orbits around a common barycenter B (center of gravity).

Both orbital ellipses with the semi-major and - minor axes a and b:

- are always aligned in the same plane (spanning a common plane)
- have different axis lengths, inversely proportional to their stellar masses
- must be similar to each other, i.e. have the same eccentricity $e = \sqrt{a^2 - b^2}/a$
- the more massive star (here M_1), always runs with the lower velocity on the smaller orbit around the barycenter
- the barycenter B is always located on the common focal point of both elliptical orbits
- M_1 and M_2 always run synchronously
- during the entire orbit the connecting line between M_1 and M_2 runs permanently through the barycenter B
- M_1 and M_2 always reach the apastron as well as the periastron at the same time.

12.2.4 Spatial Orientation of the Orbit Plane

With respect to our line of sight the orientation of the orbital planes shows a random distribution (see Figure 12.15). The angle between any axis, aligned perpendicularly to the orbital plane, and our line of sight, is defined as inclination i [49]. Thus this definition is the same for the axis inclination of stellar rotation and binary orbit planes. Analogically, $v_r \sin i$ is here the spectroscopically direct measurable share of the

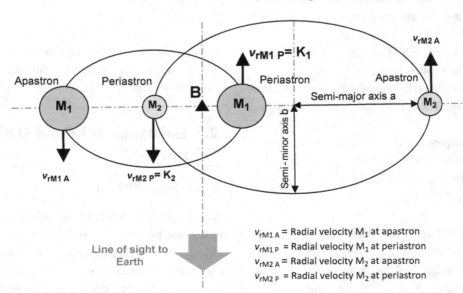

$v_{rM1\,A}$ = Radial velocity M_1 at apastron
$v_{rM1\,P}$ = Radial velocity M_1 at periastron
$v_{rM2\,A}$ = Radial velocity M_2 at apastron
$v_{rM2\,P}$ = Radial velocity M_2 at periastron

Figure 12.14 Schematic binary system

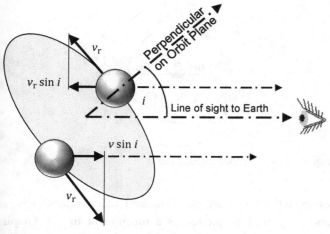

Figure 12.15 Spatial orientation of the orbit plane

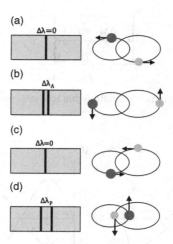

Figure 12.16 Four phases of one complete orbit

radial velocity v_r, which is projected into the line of sight to Earth. For $i = 90°$ we see the orbital ellipses exactly "edge on" i.e. $\sin i = 1$. In this case we observe an eclipsing binary [1]. For $i = 0°$ we see the orbital planes "face on" i.e. $\sin i = 0$. In this case we are unable to measure any radial velocity.

This elliptical orbit, with a given and fixed inclination i, may be rotated freely around the axis of our line of sight, without any consequences for their apparent form. Thus for circular binary orbits the inclination i fixes the only degree of freedom, which affects the apparent shape of the orbit. In contrast to the circle, the orientation of the ellipse axes in a given orbital plane forms an additional degree of freedom, determining their apparent form.

If the inclination i remains unknown, the results can be analyzed statistically only, similar to the $v \sin i$ values of the stellar rotation. Caveat: There are reputable sources defining inclination i as the angle between the line of sight and the orbital plane, similar to the convention for the inclination angle between planetary orbits and the ecliptic. The two conventions can easily be converted in to each other, applying the complementary angle $90° - i$.

12.2.5 Analysis of the Doppler Shift $\Delta\lambda$ in SB2 Systems

The distinction between SB1 and SB2 systems is explained in [1]. The diagrams in Figure 12.16 show four phases of one complete orbit. Equations {12.14} and {12.15} are derived from the spectroscopic Doppler law, Equation {11.1}. And λ_0 denotes the rest wavelength of the considered absorption line.

Phase A

The orbital velocities v_M are directed perpendicularly to our line of sight and thus the measured radial velocity yields $v_r = 0$. The spectrum remains unchanged, thus $\Delta\lambda = 0$.

Phase B

In the apastron the orbital velocities v_M reach a minimum. Directed in our line of sight they correspond to the measured radial velocities $v_r = v_M$. Thus the spectral line appears to be slightly split:

$$\Delta\lambda_A = (v_{rM1\ A} + v_{rM2\ A})\,\lambda_0/c \qquad \{12.14\}$$

Phase C

Analogously to phase A the orbital velocities v_M are directed perpendicularly to our line of sight and thus the measured radial velocity yields $v_r = 0$. The spectrum remains unchanged, thus $\Delta\lambda = 0$.

Phase D

In the periastron the orbital velocities $v_M = K$ are reaching a maximum. Directed in our line of sight they correspond to the measured radial velocities $v_r = v_M$. Thus the spectral line appears here to be strongly split.

$$\Delta\lambda_P = (v_{rM1\ P} + v_{rM2\ P})\,\lambda_0/c \qquad \{12.15\}$$

After a simple change to the Equations {12.14} and {12.15} the observed line splitting $\Delta\lambda$ allows here to compute the sum of the measured radial velocities $v_{rM1} + v_{rM2}$ for each point on the orbit.

$$v_{rM1} + v_{rM2} = (\Delta\lambda/\lambda_0)c \qquad \{12.16\}$$

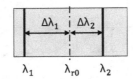

Figure 12.17 Asymmetric splitting due to a large mass difference

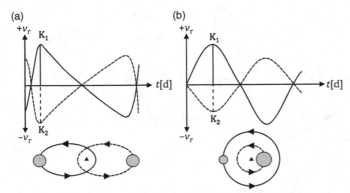

Figure 12.18 Graph (a): elliptic orbit with moderate eccentricity. Graph (b): nearly circular orbit

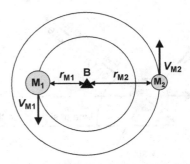

Figure 12.19 Simplified circular orbits

12.2.6 The Calculation of the Individual Radial Velocities of M_1 and M_2

If the mass difference is large enough, the splitting of the spectral line occurs asymmetrically with respect to the neutral wavelength λ_{r0} (see Figure 12.17). Applying these uneven distances $\Delta\lambda_1$ and $\Delta\lambda_2$, the individual radial velocities v_{rM1} and v_{rM2} can be calculated separately, analogously to Equation {12.15}. Anyway for this rather complex procedure a relative measurement of $\Delta\lambda$ is no longer sufficient but heliocentrically corrected measurements of the individual wavelengths λ_1 and λ_2 would be required. Additionally, due to the radial movement of the entire binary star system v_{rSyst}, the neutral rest wavelength λ_0 is shifted to λ_{r0} [49].

$$\lambda_{r0} = \lambda_0 + \frac{v_{rSyst}}{c}\lambda_0 \qquad \{12.17\}$$

If, in respect to λ_{r0}, no asymmetry of the split line occurs, M_1 and M_2 are approximately equal and the sum of the two radial velocities $v_{rM1} + v_{rM2}$ needs just to be halved.

12.2.7 The Estimation of Some Orbital Parameters in SB2 Systems

Based on purely spectroscopic observations some of the orbital parameters of the binary system can roughly be estimated. First of all, the measured radial velocities v_{rM1} and v_{rM2} must be plotted as a function of time t. Graph (a) in Figure 12.18 shows one orbital period of two stars with similar masses on elliptic orbits with moderate eccentricity (e.g. Mizar A, ζ Uma). Graph (b) displays the same but for nearly circular orbits and of stars with widely different masses. Both figures show the important maximum amplitudes K_1 and K_2. The more the velocity curves show a sinusoidal shape, the lower is the eccentricity of the orbital ellipses.

12.2.7.1 The Orbital Period T
The orbital period T can directly be determined from the course of the velocity curves. As the only variable it remains largely unaffected by any perspective effects and thus it is quite easily and accurately determinable.

12.2.7.2 Simplification to Circular Orbits
Since we are confronted with randomly oriented, elliptical orbits, a determination of orbital parameters by amateurs with reasonable accuracy is very complex. For this purpose, in addition to the spectroscopic, other types of measurements would be needed. Only for eclipsing binaries, such as Algol, already a priori a probable inclination of $i \gtrsim 80°$ can be assumed. For the rough estimation of some parameters the elliptical can be simplified to circular orbits in which radii r_M as well as orbital velocities v_M remain constant. The following considerations are all based on circular orbits. The almost unknown inclination is expressed in the equations with the term $\sin i$. For very close spectroscopic binaries with orbital periods of just a few days and rather small eccentricities of the orbit ellipses, the error due to this simplification is anyway limited (see Figure 12.19).

12.2.7.3 The Orbital Velocity v_M
To determine the orbital velocity v_M we need from the velocity diagram the maximum v_{rM} values for both components.

By definition they correspond to the maximum amplitudes K_1 and K_2. For the circular orbit velocity v_M follows:

$$K = v_M \sin i \quad \text{or} \quad v_M = \frac{K}{\sin i} \qquad \{12.18\}$$

12.2.7.4 Estimation of the orbital radii r_M

With the orbital period T and the velocity $v_M = K / \sin i$ for a circular orbit the corresponding radius r_M can be calculated. From geometric reasons it generally follows that:

$$T = \frac{2 r_M \pi}{K} \sin i \qquad \{12.19\}$$

$$r_M = \frac{KT}{2\pi \sin i} \qquad \{12.20\}$$

If both lines can be examined, the obtained values for K_1 and K_2 allow to compute separately the corresponding radii r_{M1} and r_{M2}.

12.2.7.5 Estimation of the Sum of Stellar Masses $M_1 + M_2$ in SB2 Systems

If both lines can be analyzed, then with Equation {12.22} the total mass of the system M_1+M_2 can be determined. Equation {12.22} is obtained if for simplified circular orbits, the sum of the partial radii r_{M1} and r_{M2}, expressed according to Equation {12.20}, is inserted into Equation {12.21} as "semi-major axis" $a = r_{M1} + r_{M2}$. This equation combines the Newtonian gravitation with Kepler's third law:

$$M_1 + M_2 = \frac{4\pi^2 a^3}{GT^2} \qquad \{12.21\}$$

$$M_1 + M_2 = \frac{T(K_1 + K_2)^3}{2\pi \, G \sin^3 i} \qquad \{12.22\}$$

Where the gravitational constant $G = 6.674 \times 10^{-11} \, \text{m}^3 \, \text{kg}^{-1} \, \text{s}^{-2}$. The partial masses can then be estimated by their total mass M_1+M_2, applying the partial radii r_{M1} and r_{M2}:

$$M_1 r_{M1} = M_2 \, r_{M2} \qquad \{12.23\}$$

$$M_1 = \frac{r_{M2}}{r_{M1} + r_{M2}} \, (M_1 + M_2)$$

$$M_2 = \frac{r_{M1}}{r_{M1} + r_{M2}} \, (M_1 + M_2) \qquad \{12.24\}$$

For binary star systems M is often expressed in solar masses $M_\odot$ and the distance in astronomical units (AU). For conversion: 1 solar mass $M_\odot \approx 1.98 \times 10^{30}$ kg and 1 AU $\approx 149.6 \times 10^9$ m.

12.2.7.6 Estimation of the Stellar Masses M_1 and M_2 in SB1 Systems

The limitation of the analysis to only one line has of course negative consequences on the information content and accuracy of the system parameters to be determined. In spectra of SB1 systems, only the radial velocity curve of the brighter component M_1 can be examined and therefore, aside from the orbital period T, just the amplitude K_1 can be determined. So with Equation {12.20}, only the corresponding orbital radius r_{M1} can be estimated. Unfortunately, neither the total mass M_1+M_2 nor the partial masses M_1, M_2 are directly determinable. These variables are also found on the left side of the so-called "mass function" (Equation {12.25}). On the other side of this equation we see all measurable or known items like the orbital period T, the amplitude K_1 and the gravitational constant G:

$$f(M): \quad \frac{(M_2 \sin i)^3}{(M_1 + M_2)^2} = \frac{TK_1^3}{2\pi G} \qquad \{12.25\}$$

For stellar physics this equation only becomes helpful if we can estimate one of the masses, for example by the spectral class. In this case it provides for M_2 a minimal mass and the result remains loaded by the uncertainty of the mostly unknown inclination $\sin^3 i$. The mass function plays an important role, for example, for the mass estimation of black holes and extrasolar planets.

13 Gravity, Magnetic Fields and Element Abundance

13.1 Measurement of the Surface Gravity

13.1.1 Overview

The surface gravity of a star is the gravitational acceleration which a hypothetical test particle with infinitesimal low mass experiences at the star's surface. It is calculated as the decimal logarithm of the gravity g, expressed in cgs units of $[cm\ s^{-2}]$.

If the radius R and mass M of the star are known it is possible to calculate the surface gravity by Equation {13.1}:

$$g = G\ \frac{M}{R^2} \qquad \{13.1\}$$

where G is the universal gravitational constant. Expressed in solar units ($M_\odot$, $R_\odot$) and transformed in logarithmic form this finally yields Equation {13.2}:

$$\log g = \log M - 2 \log R + 4.437 \qquad \{13.2\}$$

In case of binary star systems it is possible to determine M and R, but for the majority of stars a direct measurement is not applicable. Therefore other indirect methods are necessary. For example, the Sun with $M = 1$ and $R = 1$, Equation {13.2} yields $\log g = 4.437$.

13.1.2 Method Based on the Wilson–Bappu Effect

The Wilson–Bappu effect, already presented in Section 10.3.4, also offers an option to estimate the surface gravity of a giant within the spectral classes ~G–M. Figure 10.9 shows how the width W of the reversal emission core of the Ca II K line ($\lambda 3933.66$) is measured. However, considered in this context,

W is increasing here by decreasing surface gravity. Based on atmospheric model calculations the following relationship, expressed by Equation {13.3}, has been proposed between the width W of the emission reversal, the surface gravity and T_{eff} [57]. It is based on measurements by an echelle spectrograph with a resolution R ~45,000, attached to the 1.8-m optical telescope at Bohyunsan Optical Astronomy Observatory (BOAO, South Korea). During 2008–2012 some 70 stars from spectral classes G–M were analyzed in the wavelength range of 3600–10,500 Å. Further 53 echelle spectra of 53 late-type stars (G, K and M) were adopted from UVES POP with R ~80,000.

$$\log g = -5.85 \log W + 9.97 \log T_{\text{eff}} - 23.48 \qquad \{13.3\}$$

For example, for Arcturus in Section 10.3.4 it was already determined to be $W \approx 74.8$ km s^{-1}. Assuming T_{eff} to be 4300 K, Equation {13.3} yields for the surface gravity $\log g \approx 1.7$. Depending on the source the published $\log g$ values for α Boo are roughly within a range of 1.5–1.7.

13.1.3 Further Surface Gravity Indicators

A star with typically one solar mass starts its life as a relatively dense dwarf on the main sequence with a surface gravity $\log g \approx 4.4$. Later on its radius expands dramatically to that of a red giant, exhibiting an atmosphere of very low density and a typical $\log g < 3.5$, causing at this stage a dramatic loss of mass. Finally, the star ends as an extremely small and dense white dwarf with a surface gravity of $\log g > 7$. This huge covered range comprises several orders of magnitude, reached by each star during its life with an initial mass

$M_i \lesssim 8\ M_\odot$, allowing in this way an estimation of its observed state of development.

For rather hot stars with $T_{eff} > 7000$ K, and mainly caused by the Stark effect [7], the FWHM values of photospheric hydrogen are highly pressure sensitive. The dramatic slimming of the H lines with decreasing density (luminosity effect) is demonstrated in the *Spectral Atlas* [1] by a montage of profiles by the sequence Vega (A0V), δ Cas (A5III-IV) and Deneb (A2Ia). These hot stars have similar spectral classes, but very different luminosities. Vice versa the tremendous increase of FWHM with rising surface gravity is shown again in [1] by a montage of Sirius (A1Vm) and the white dwarf 40 Eridani B (DA2.9).

However, for rather cool stars with $T_{eff} < 7000$ K this strong dependency of FWHM from the surface gravity gets increasingly lost. This is impressively demonstrated in [1] by a montage of three F class stars with different luminosities, exhibiting no visible differences at FWHM of the H lines. However, in this case another method comes in to play using the influence of the surface gravity on the ionization equilibrium, mainly observable at certain sensitive Fe and Ti lines. The use of this spectroscopic indicator can be demonstrated by the following simplified approach. According to the astronomical notation system a neutral metal atom is represented here as (X)I and the ionized form as (X)II. The ionization equilibrium for this element is defined by Equation {13.4}:

$$(X)I \rightleftarrows (X)II + e^- \qquad \{13.4\}$$

where (X)I represents the starting position as a neutral metal atom, (X)II the ionized form or next ionization stage and e^- represents the electron. Depending on the electron density two possibilities can occur.

- A high surface gravity compresses the stellar photosphere. As a result the distance between the atoms, ions and electrons decreases, so the relative electron density increases. To restore the ionization equilibrium the relative "excess" of electrons is eliminated by recombination. Therefore the equilibrium of Equation {13.4} is pushed to the left, generating less ionization.
- In case of lower surface gravity the stellar photosphere is more decompressed. As a result more distance is created between the atoms, ions and electrons, so the electron density decreases. To restore here the equilibrium the relative "shortage" of electrons is eliminated by ionization. Therefore the equilibrium of Equation {13.4} is pushed to the right, generating more ions and electrons.

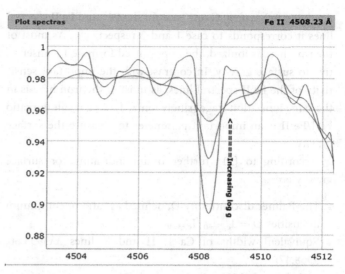

Figure 13.1 Synthetic spectral lines of Fe II λ4508.23 showing surface gravity dependencem, made by VO-SPECFLOW-POLLUX (Marc Trypsteen)

Typical examples are the spectral lines of the Fe I and Fe II lines [7]. Applying Equation {13.4} yields Fe I $\rightleftarrows$ Fe II + e^-. The electronic configuration of Fe I is indicated in Table 3.2. Apart from the closed sub shells the important energy levels are the $4s^2$ and the $3d^6$. Removing one electron from the $4s^2$ shell generates single ionized iron Fe II and by removing the second one from the same shell double ionized iron is generated known as Fe III. As a result of all possible transitions within and from the open $3d^6$ shell, inclusive of the $4s$ level, numerous spectral lines from Fe I and Fe II emerge in the ultraviolet, optical and infrared wavelength region, making them important indicators to determine stellar parameters [12].

The effect of surface gravity on the ionization equilibrium and therefore on the spectral lines are demonstrated with the synthetic spectrum of a K type star zoomed in on the Fe II line at λ4508.23 as shown in Figure 13.1.

In this case of a rather cool star ($T < 7000$ K) the pressure dependency of specific spectral lines can be predicted by rules depending on the ionization stage according to [6] and [21]:

1. Most lines of this element represent the next higher ionization stage. In that case the considered line is insensitive to surface gravity changes.
2. Most lines of this element represent the same ionization stage. In that case the considered line is sensitive to surface gravity changes.
3. Most lines of the element represent the next lower ionization stage. In that case the considered line is highly sensitive to surface gravity changes.

Applying to a type F, G or K star for the Fe I and Fe II spectral lines it corresponds to case 1 and 2 respectively. As most of the iron here is ionized, the non-ionized form Fe I is insensitive to surface gravity. In contrast with the important sensitivity of the Fe II spectral lines, as most of the iron here is in the same ionization stage (singly ionized). As a result the ratio Fe I/Fe II is an important parameter to measure the surface gravity.

According to [21] further useful indicators for surface gravity are:

- The Balmer discontinuity D, defined as ratio of continuum intensities $D = I_{c\lambda3660}/I_{c\lambda3640}$
- Equivalent widths of Ca II H and K lines $\lambda\lambda3933.66$, 3968.47:
  - Ca I line $\lambda6162$
  - Na I D lines $\lambda\lambda5889.95, 5895.92$
  - Mg I b lines $\lambda\lambda5183.62, 5172.70$ and 5167.33 (magnesium triplet).

13.2 Measurement of Stellar Magnetic Fields

13.2.1 Overview

For astronomical objects the Zeeman effect enables a rough estimation of the magnetic flux density. On the level of the magnetic quantum number m it generates a split up of the energy levels, which consequently results in a split of individual spectral lines (Section 3.2.1). The Zeeman effect is from great importance for the measurement of stellar magnetic fields, as well as for a detailed magnetic survey of the solar surface. With high-resolution spectrographs amateurs can observe this effect mainly qualitatively and for a few stellar objects even the rough estimation of the magnetic flux density B is feasible. Here follows a strongly simplified presentation of this rather complex theory.

Analogously acts the so-called Stark effect for electric fields. It plays, among others, a major role by the broadening of the H-Balmer lines, impressively recognizable in spectra of white dwarfs. Otherwise, for amateurs this principle offers no further significant observation opportunities.

13.2.2 The Zeeman Effect

This effect was discovered in 1896 by the Dutch Nobel Prize winner Pieter Zeeman. If an external magnetic field with flux density B acts on the mechanical angular momentum and the

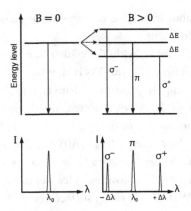

Figure 13.2 Normal Zeeman effect

magnetic dipole moment of an orbiting electron, the corresponding energy level gets symmetrically split up by the amount of $2 \times \Delta E$. Thereby the accordingly generated emission or absorption components get polarized differently.

Figure 13.2 shows the so-called "normal Zeeman effect" with a neutrally remaining π-component and two secondary lines $\sigma+$ and $\sigma-$, each shifted by $\Delta\lambda$. This applies for the so-called "transverse effect" where the field lines run perpendicularly to our line of sight. In the case of the "longitudinal effect" they run parallel. In this case, just the two shifted σ components remain and the neutral π-component disappears. In special cases the "anomalous Zeeman effect" may split some lines even into more than three components. The shift amounts of ΔE and $\Delta\lambda$ are proportional to the magnetic flux density B. This allows a simple calculation of B, applying Equation {13.5}:

$$B = \frac{\Delta\lambda}{4.67 \times 10^{-13} g_{eff} \lambda^2} \qquad \{13.5\}$$

Here, λ is the wavelength in [Å], and B the magnetic flux density in [G]. In astrophysics the unit gauss [G] is still in use, i.e. 10 kG [kilogauss] $\triangleq$ 1 T [tesla]. The dimensionless, gyromagnetic or "effective Landé factor" g_{eff}, compensates for the different spreads $\Delta\lambda$, caused by the unequal magnetic sensitivities of the individual spectral lines (Alfred Landé, German–American physicist 1888–1976). It is determined experimentally because the possible theoretical calculations, based on the corresponding electron transitions, normally display too large deviations [28]. For example, calculated for the Fe I line at $\lambda6302.5$ with a Landé factor of $g_{eff} = 2.5$ and an assumed magnetic flux density of $B \approx 10$ kG, Equation {13.5} yields $\Delta\lambda \approx 0.46$ Å. However, the line is being split by twice the amount $2 \times \Delta\lambda$. It is further important to note that $\Delta\lambda$ increases by the square of the wavelength. Thus for

Table 13.1 Strong Landé factors for absorptions appearing in the solar spectrum

Line	λ	g_{eff}	Line	λ	g_{eff}	Line	λ	g_{eff}
Cr I	4654.724	3	Cr I	5247.565	2.5	Fe I	6082.708	2
Fe I	4704.948	2.5	Fe I	5250.208	3	Ca I	6102.723	2
Fe I	4872.136	2.3	Fe I	5254.955	2.3	Fe I	6173.334	2.5
Fe I	4878.208	3	Cr I	5264.153	2	Fe I	6213.429	2
Fe I	4938.813	2	Fe I	5497.516	2.3	V I	6258.571	3.3
Fe I	5131.468	2.5	Fe I	5506.778	2	Fe I	6302.494	2.5
Fe I	5191.463	2	Fe I	5724.454	2.4	Fe I	6733.157	2.5
Fe I	5226.862	2.2	Fe I	5807.782	3			

measurement purposes unblended long-wave lines with an additionally high magnetic sensitivity g_{eff} are preferred.

13.2.3 Spectral Lines with Strong Landé Factors

For spectral lines with strong Landé factors, generally the following applies:

$g_{\text{eff}} = 0$: No split up of the line occurs, see Equation {13.5}
$g_{\text{eff}} = 1$: Magnetically non-sensitive line
$g_{\text{eff}} = 2$: Magnetically sensitive line
$g_{\text{eff}} = 3$: Magnetically increased sensitive line.

Table 13.1 displays for the solar spectrum (G2V) some absorptions with Landé factors $g_{\text{eff}} \gtrsim 2$, extracted from Spectroweb [23].

13.2.4 Possible Applications for Amateurs

Spectrographically no details can be resolved on stellar surfaces. That is why the variable B corresponds here to a mean magnetic field, which has been averaged over the entire visible hemisphere of the star, also called "mean magnetic field modulus" [29]. This simple, so called "integrated field measurement" [30], by exclusive application of $\Delta\lambda$, is even feasible for accordingly equipped amateurs. A practical example can be found in [1], with Babcock's Star.

The determination of the magnetic vector field by the so-called Stokes parameters requires in addition a polarimetric investigation, which is feasible only by a few specialized amateurs. However, this method is particularly important for professional solar research, to analyze the magnetic fields of small-scale objects, such as sunspots. If such features are very large, they may reach field strengths of up to 3 kG. In this case, with a resolution power of R $\gtrsim$ 20,000 within strong sunspots even amateurs may observe at least a relative increase of the line width. In the integrated light of a star, a direct comparison with a magnetically just slightly disturbed line is missing. Therefore, in the amateur domain a clearly recognizable split up is required. However, this is met only by very few stars of the Ap class, exhibiting only slightly rotationally broadened lines $v \sin i \lesssim 10$ km s^{-1} but very strong magnetic fields $B \gtrsim 10$ kG, whose flux density is mostly variable. A list of such stars is found in [1].

13.3 Abundance of Elements

13.3.1 Astrophysical Definition of Element Abundance

In astrophysics, the abundance of an element A_{el} is expressed as decimal logarithm of the ratio between its amount of particles per unit volume N_{el}, and those of hydrogen N_{H}, whose abundance is defined to $A_{\text{H}} = 12$ [16] (the mass ratios do not matter here):

$$A_{\text{el}} = \log_{10} \frac{N_{\text{el}}}{N_{\text{H}}} + 12 \qquad \{13.6\}$$

By transforming logarithmically we directly obtain the relationship $N_{\text{el}}/N_{\text{H}}$:

$$\frac{N_{\text{el}}}{N_{\text{H}}} = 10^{(A_{\text{el}}-12)} \qquad \{13.7\}$$

13.3.2 Astrophysical Definition of Metallicity Z (Metal Abundance)

Of great importance is the ratio of iron to hydrogen $N_{\text{Fe}}/N_{\text{H}}$. This is also computed with the relative number of atoms per unit volume and not with their individual masses. The metallicity Z in a stellar atmosphere, also called [Fe/H], is expressed as decimal logarithm in relation to the Sun:

$$Z = [\text{Fe/H}] = \log_{10} \frac{(N_{\text{Fe}}/N_{\text{H}})_{\text{star}}}{(N_{\text{Fe}}/N_{\text{H}})_{\text{Sun}}} \qquad \{13.8\}$$

Metallicity Z values smaller than found in the atmosphere of the Sun are considered to be metal poor, carrying a negative sign (–). The existing range actually reaches from approximately +0.5 to –5.4. Iron is used here as representative of the metals because it appears quite frequently in the spectral profile and is relatively easy to analyze.

As outlined in the *Spectral Atlas* [1] the Z value allows estimation of the age of a star cluster. The most efficient way, has proven to be the analysis of the Ca II calcium triplet (CaT) at λλ8542, 8498 and 8662. Anyway the near infrared range is somewhat outside the reach of most amateur equipment and frequently contaminated by blends with other absorptions.

13.3.3 Quantitative Determination of the Abundance

The identified spectral lines of the examined object inform directly which:

- elements and molecules are present
- isotopes of an element are present (restricted to some cases and to highly resolved profiles)
- stages of ionization are generated.

In this context the quantitative determination of the abundance can be outlined here just very roughly. It is highly complex and cannot be obtained directly from the spectrum. It requires additional information, which can be obtained only with simulations of the stellar photosphere [6]. The intensity of a spectral line is an indicator which provides information on the abundance of a particular element. However, this value is further influenced, for example, by the effective temperature T_{eff}, the pressure, the gravitational acceleration, as well as the macro-turbulence and the rotational speed of the stellar photosphere. Furthermore, T_{eff} also affects the degree of ionization of the elements, which can be estimated by the Saha equation [6].

These complications are impressively demonstrated in the solar spectrum. Over 90% of the solar photosphere consists of hydrogen atoms with the defined abundance of $A_{\text{H}} = 12$. Nevertheless, as a result of a too low temperature of ~5800 K, the intensity of the H-Balmer series remains remarkably modest. The dominating main features of the solar spectrum, however, are the two Fraunhofer H and K lines of ionized calcium Ca II , although its abundance is just $A_{\text{Ca}} \approx 6.36$ [3]. However, from quantum mechanical reasons, at solar photospheric temperature of 5800 K, Ca II is an extremely effective absorber and the optimum conditions for the hydrogen lines are reached not until ~10,000 K. In the professional area the element abundance is also determined by iterative comparison of the spectrum with simulated synthetic profiles calculated for different chemical compositions [6].

13.3.4 Relative Abundance of Stars of Similar Spectral Class

A simplified special case is formed by stars with similar spectral and luminosity class and comparable rotational velocities. Thus, the physical parameters of the photospheres are very similar. Here the equivalent widths of certain lines can simply be compared and thus the relative abundance differences at least qualitatively be seen. In [1] this effect is demonstrated at the classical example of the two main sequence stars Sirius A1Vm and Vega A0V. Figure 13.3 shows the basic principle, the so called "curve of growth."

The curve of growth describes the behavior of the EW of an absorption (or emission) line in relation to the number of atoms, which are involved in their generation. At the bottom left side of Figure 13.3 we see the unsaturated part of the model curve where the gas is optically thin. This results here in a somewhat linear run of the curve. For amateurs this section is interesting for an easy analysis of the abundance because the number of atoms is here nearly proportional to the equivalent width ($W_\lambda \propto fN$). Adding even more atoms

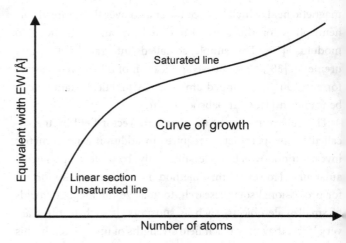

Figure 13.3 Curve of growth

the gas becomes optically thick and the curve of growth reaches the saturation, running now more horizontally, what is called the saturation plateau phase $\left(W_\lambda \propto \sqrt{\ln\left(fN\right)}\right)$. This and the following behavior is also impressively demonstrated by comparison with Figure 9.8. Adding now still more atoms increases the density drastically but the EW grows less pronounced here by the limited contribution coming just from the outermost parts of the wings on both sides of the line $\left(W_\lambda \propto \sqrt{fN}\right)$. This way the model curve finally reaches here the upper right end [104].

14 Analysis of Emission Nebulae

14.1 The Balmer Decrement

14.1.1 Introduction

In spectra of emission nebulae, exhibiting the H Balmer series in emission, the line intensity is fading with decreasing wavelength. It is highest at the Hα line and gets continuously weaker towards shorter wavelengths Hα > Hβ > Hγ > Hδ etc. This phenomenon is called Balmer decrement, D. In the very distant outskirts of H II regions and planetary nebulae the theoretical Balmer decrement can be observed in its purest form because the emission lines are generated here almost exclusively by recombination of ionized hydrogen. The high-energy photons above the Lyman limit, necessary for the ionization, are generated here by an extremely hot central star, which specific spectral radiation distribution has otherwise no significant impact on the decrement values. Thus, the intensity of the H Balmer series is fading reproducibly, following the quantum mechanical laws. This effect is a highly important indicator for astrophysics.

14.1.2 Definition of the Balmer Decrement

For most astrophysical analyses the measured intensity ratio of the Hα and Hβ lines D_{obs} is required, which is formed by the relative Hα and Hβ fluxes, according to Equation {14.1}. This corresponds to the definition of the Balmer decrement:

$$D_{obs} = F(H\alpha)/F(H\beta) \qquad \{14.1\}$$

By convention, the intensity values are related to Hβ = 1, sometimes also to Hβ = 100.

14.1.3 Theoretical Balmer Decrement for Emission Nebulae

Table 14.1 shows the theoretical decrement values D_{Th}, by Brocklehurst [52], quantum mechanically calculated for gases with a low as well as for such with a high electron density N_e and electron temperatures T_e = 10,000 K and 20,000 K.

In expert jargon the two density cases are called "Case A" and "Case B."

Case A: A very thin gas, which is transparent for Lyman photons, so they are able to escape the nebula.

Case B: A dense gas, which retains the short-wavelength ($\lambda <$ Lyman α)photons.

The differences are relatively modest. For analyses in the field of H II regions and planetary nebulae, the theoretical decrement D_{Th} = 2.85 is widely established today.

Table 14.1 Theoretical Balmer decrement for cases A and B

H Line	Case A $D_{Th} = F(H)/F(H\beta)$ Low Density ($N_e = 10^2$ cm^{-3})		Case B $D_{Th} = F(H)/F(H\beta)$ High Density ($N_e = 10^6$ cm^{-3})	
	T_e =10,000 K	T_e =20,000 K	T_e =10,000 K	T_e =20,000 K
Hα	2.85	2.8	2.74	2.72
Hβ	1	1	1	1
Hγ	0.47	0.47	0.48	0.48
Hδ	0.26	0.26	0.26	0.27

14.1.4 Balmer Decrements at Stellar and other Astronomical Objects

Stellar objects generate hydrogen emissions under considerably more complex and turbulent conditions which are influenced also by the radiation characteristics of the star. The decrement may run differently steep here or even inversely. Figure 14.1 displays the different cases according to the terminology of T. Kogure [22] and in comparison, the steep Balmer decrement $D_{obs} \approx 5$ of the LBV star P Cygni, relative to a unified radiation intensity $I_C = 1$.

In the *Spectral Atlas* [1] the following practical examples can be considered:

Steep: LBV star P Cygni, protostar T Tauri, protostar R Monocerotis, starburst galaxy M 82, Seyfert galaxy M77, semi-detached binary β Lyrae
Nebular: H II regions and planetary nebulae
Slow: Quasar 3C273

Inverted: Flare star Gliese 388, LBV star Mira, S-type star R Cygni
Irregular: Nova Delphini, recurrent nova T Crb, dwarf nova SS Cygni.

14.1.5 Applications of the Balmer Decrement in the Amateur Sector

At the amateur sector, the application focuses on spectra of emission nebulae. Here, mainly the ratio of the measured to the theoretical Balmer decrement is evaluated. This allows, for example, estimation of the extinction by interstellar matter $D_{ISM}(\lambda)$. Conversely, the recorded emission lines can be dereddened if compared to the theoretical Balmer decrement. At stellar objects, the measured Balmer decrement may be a classification criterion, so at PMS protostars, analyzing the attenuation $D_{CSM}(\lambda)$ by circumstellar dust [1] [7].

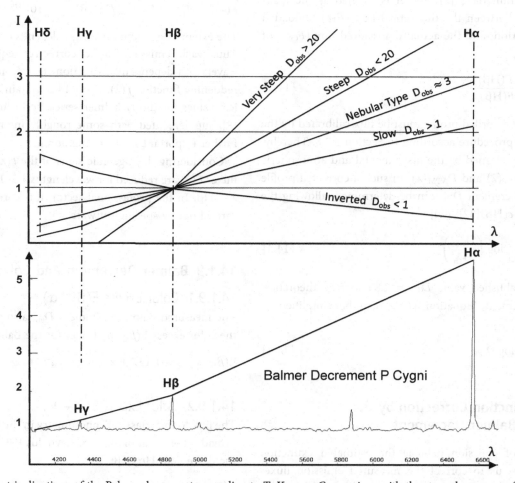

Figure 14.1 Different inclinations of the Balmer decrement, according to T. Kogure. Comparison with the steep decrement of P Cygni.

14.1.6 Measurement of the Balmer Decrement by Amateurs

The very slim lines in the spectra of emission nebulae usually appear superposed to a barely noticeable, diffuse continuum. In this case the continuum-related EW values, representing the relative flux, turned out as not suitable and further remain unadjusted by a proportional flux calibration according to Section 8.2. For amateurs, as a reasonable approximation, it is often easier to measure the peak intensities by the relative length I_E above the continuum level, instead of determining the fluxes F_E by intergrating the profile areas of the emissions.

14.1.7 Spectroscopic Estimation of Interstellar Extinction

For spectra of emission nebulae the extinction parameter c comprises the entire attenuation along the total line of sight between the observed object and the outer edge of Earth's atmosphere (see Section 8.2.2). It is called the extinction parameter c [8], and it is defined as the logarithmic ratio between the theoretical flux (F_{Th}), unloaded of any extinction, and the actually measured flux (F_{Obs}) of the $H\beta$ line:

$$c(H\beta) = \log \frac{F(H\beta)_{Th}}{F(H\beta)_{Obs}} \qquad \{14.2\}$$

First of all the recorded profile must be flux calibrated by the standard star procedure according to Section 8.2.8. Thereby, it gets roughly cleaned by the instrumental and atmospheric responses $D_{INST}(\lambda)$ and $D_{ATM}(\lambda)$. In such a corrected profile the Balmer decrement D_{obs} can be determined, allowing the estimation of $c(H\beta)$[32], [50]:

$$c(H\beta) = \frac{1}{0.325} \log \left(\frac{D_{Obs}}{D_{Th}} \right) \qquad \{14.3\}$$

With the established value $D_{Th} = 2.85$ for the theoretical Balmer decrement, Equation {14.3} can be simplified as follows:

$$c(H\beta) = 3.1 \cdot \log D_{Obs} - 1.4 \qquad \{14.4\}$$

14.1.8 Extinction Correction by the Measured Balmer Decrement

For spectra of emission nebulae the estimated extinction $c(H\beta)$ enables us to correct the measured emission fluxes in respect of interstellar extinction. With the following

Table 14.2 $f(\lambda)$-values for important nebular emissions

Element/λ	$f(\lambda)$
S II 6732	−0.360
Hα 6563	−0.334
He I 5876	−0.23
O III 5007	−0.07
Hβ 4861	0
He II 4686	0.02
Hγ 4340	0.114
Hδ 4102	0.168
Hε 3970	0.204

correction function, the fluxes of the emission lines $F(\lambda)$ are adjusted (or "dereddened") relatively to $F(H\beta)$[8]:

$$F(\lambda)/F(H\beta) = F(\lambda)_{Obs}/F(H\beta)_{Obs} \times 10^{c(H\beta)f(\lambda)} \qquad \{14.5\}$$

The extent of extinction strongly depends on the wavelength. Thus each emission to be corrected requires a specific, wavelength-dependent extinction factor, provided by the reddening function $f(\lambda)$. Table 14.2 contains for $f(\lambda)$ numerical values for the H Balmer series, according to Gurzadyan [8], supplemented with some roughly interpolated data for further important nebular emission lines.

Considering the algebraic signs of the $f(\lambda)$ values, the line fluxes $F(\lambda)$ are reduced for wavelengths $\lambda > H\beta$ and raised for $\lambda < H\beta$. In the professional sector, this correction is usually carried out by specific software.

14.1.9 Balmer Decrement and Color Excess

14.1.9.1 Color Excess E($\beta - \alpha$)

The measured Balmer decrement D_{obs} also allows estimating the color excess E($\beta - \alpha$) (mag) for the Balmer lines [8]:

$$E(\beta - \alpha) = -1.137 + 2.5 \cdot \log D_{Obs} \quad [mag] \qquad \{14.6\}$$

14.1.9.2 Color Excess E($B - V$)

The link to the "classical" photometry in the E($B - V$) system provides the equation of C. S. Reynolds [53], transformed for the decrement Hα/Hβ:

$$E(B - V) = 2.21 \cdot \log D_{Obs} - 0.975 \quad [mag] \qquad \{14.7\}$$

14.2 Plasma Diagnostics for Emission Nebulae

14.2.1 Preliminary Remarks

In the *Spectral Atlas* [1] a classification system is presented for the excitation classes of the ionized plasma in emission nebulae. Further, this process is practically demonstrated based on several objects. Here additional information to the physical background and further diagnostic possibilities are presented, based, among others, on [8], [31], [54]. At first these processes are schematically demonstrated by a hydrogen atom.

14.2.2 The Photoionization in Emission Nebulae

The high-energy UV photons from the central star ionize the atoms of the surrounding nebula and thus get completely absorbed. Therefore, at the latest the outside of the so-called Strömgren sphere ends the partially ionized plasma of the emission nebula. Since the observed intensity of spectral lines barely show any fluctuations, a permanent equilibrium between newly ionized and recombined ions must exist (see Figure 14.7 later). Rough indicators for the strength of the radiation field are the species, the ionization stage and the abundance of the generated ions. The first two parameters can be derived directly from the spectrum and compared with the required ionization energy.

14.2.3 Kinetic Energy and Maxwellian Velocity Distribution of Electrons

The free electrons, released by the ionization process, form an electron gas (Figure 14.2). Their kinetic energy E_e corresponds to the surplus energy of the UV photons, heating up

the particles of the nebula. This results in a Maxwellian velocity distribution for the electrons. The electron temperature T_e[K] and density N_e[cm^{-3}] affect the following recombination and collision excitation processes. Here, T_e is directly proportional to the average kinetic energy E_e of the free electrons:

$$E_e = k_B T_e \qquad \{14.8\}$$

where the Boltzmann constant $k_B = 8.6 \times 10^{-5}$ eV K^{-1}.

The classic equation for the kinetic energy {14.9} yields E_e in joules [J] with the electron mass $m_e = 9.1 \times 10^{-31}$ kg and the average velocity v [m s^{-1}]. The short equation {14.10} yields E_e directly in electronvolts [eV]:

$$E_e = \frac{1}{2} m_e v^2 \quad [J] \qquad \{14.9\}$$

$$E_e = 2.84 \times 10^{-12} v^2 \quad [eV] \qquad \{14.10\}$$

The chart in Figure 14.3 shows schematically after Gieseking [31], the Maxwellian distribution of electron velocities displayed for relatively "cool" and "hot" nebulae, calculated for T_e 10,000 K and 20,000 K. The maximum values of the two curves correspond to the average kinetic energy according to Equation {14.8} (0.86 eV and 1.72 eV). The upper edge of the diagram is supplemented here with the values of the kinetic electron energy. Additionally mapped are further the two minimum rates, required for the excitation of the [O III] lines.

14.2.4 Significant Processes within Ionized Nebular Plasmas

Figure 14.4 shows the four most significant processes occurring within ionized nebular plasmas, schematically demonstrated with segments of a hydrogen atom.

14.2.5 Recombination Process (Case A)

If an electron hits the ion centrally, it is captured and first ends up mostly on one of the upper excited levels. The energy E_p generated this way is emitted as a photon. It corresponds to the sum of the previous kinetic energy of the electron E_e and the discrete energy difference ΔE_n due to the distance of the occupied shell to the ionization limit. Since the share of the kinetic energy E_e varies widely, from this first stage of the recombination process not a discrete but just a broadband radiation is contributed to the weak continuum:

$$E_p = \frac{1}{2} m_e v^2 + \Delta E_n \qquad \{14.11\}$$

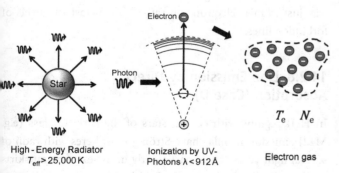

High - Energy Radiator
$T_{eff} > 25,000$ K

Ionization by UV-Photons $\lambda < 912$ Å

Electron gas

Figure 14.2 High energy photons ionize the atoms of the surrounding nebula. The released electrons finally form an electron gas with the temperature T_e and density N_e.

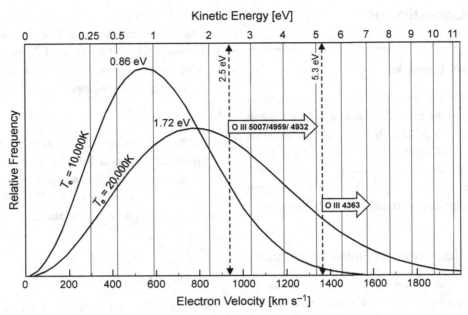

Figure 14.3 Maxwellian distribution of electron velocities displayed for nebulae with electron temperatures of $T_e = 10{,}000$ K and $T_e = 20{,}000$ K [31]

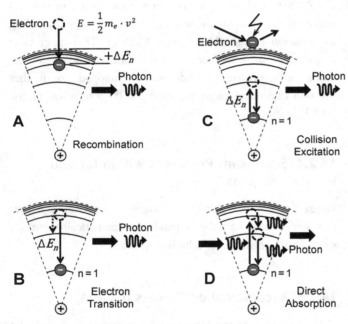

Figure 14.4 Significant processes within ionized nebular plasmas

cameras! This process cools the nebula, because each escaping photon removes energy, contributing thereby to a thermal balance with the heating process by the electron gas. This regulates the electron temperature T_e in the nebula in a range of about 5,000 K to 20,000 K.

14.2.7 Line Emission by Collision Excitation (Case C)

If an electron hits an ion then much more frequently a collision excitation occurs rather than a recombination, whereby the electron transfers a part of its kinetic energy to the ion. If the impact energy is $\geqq \Delta E_n$ the electron is briefly raised to a higher level. By allowed transitions it would immediately fall back to the ground state (within ~10^{-8} s). Anyway the prevailing temperatures in nebulae enable this way just to raise electrons on rather low metastable levels of forbidden lines.

14.2.6 Line Emission by Electron Transitions (Case B)

After recombination the electron "drops" either directly or via several intermediate levels n (cascade), to the lowest energy ground state $n = 1$. Such transitions generate discrete line emission, according to the energy difference ΔE_n. Most of these relatively low energetic photons now leave the nebula freely – including those ending on the sensor chip of our

14.2.8 Line Emission by Direct Absorption (Case D)

In H II regions with central stars of the early O class (e.g. M42) emission nebulae have Strömgren spheres with radii of several light years. Thus, particularly in the extreme outskirts of the nebula, the radiation field of the central star gets extremely diluted and the probability that the energy of a photon exactly fits to the excitation level of a hydrogen atom becomes extremely small. Therefore direct or resonance absorption of a photon does not significantly contribute to the line emission [31]. Further, the main part of the photons is radiated in the UV range. Consequently many atoms are immediately ionized once the energy of the incident photons is above the ionization limit. Therefore a substantial line emission of permitted transitions is only possible by the recombination process. The high spectral intensity of hydrogen and helium, the main actors of the permitted transitions, is caused by the abundance which is by several orders of magnitude higher than the remaining elements in the nebula. The incidence of a specific electron transition also determines the relative intensity of the corresponding spectral line.

14.2.9 Line Emission by Forbidden Transitions

Emission nebulae contain various kinds of metal ions, most of them with several valence electrons on the outer shell causing electric and magnetic interactions, which multiplies the possible energy states. Corresponding term schemes (or Grotrian diagrams) are therefore highly complex and contain also so-called "forbidden transitions." However, the extremely low density of the nebulae provides ideal conditions for the highly impact-sensitive metastable states, lasting from some seconds

up to several minutes. After this time a so called *radiative de-excitation* spontaneously occurs, generating a forbidden emission line. In the very rare case if the metastable state untimely becomes destroyed by a passing electron a so called *collision de-excitation* takes place, without emitting any radiation [31].

But first of all, these metal atoms must be ionized to the corresponding stage, which requires high-energy UV-photons. Some of the needed energies are listed in Table 14.3, compared to hydrogen and helium [eV, λ]. The higher the required ionization energy, the closer to the star the ions are generated, what is called "stratification" [8].

Once the metal atoms are ionized to the according stage, just a few additional electronvolts [eV] are required to populate the metastable initial states for forbidden transitions. This small amount of energy is plentifully supplied by the frequent collision excitations by free electrons! This explains the strong intensity of the forbidden as compared with the allowed transitions. Therefore, in the context of model computations, these metal ions are also called "cooler." Influenced by the highly effective line emission they significantly contribute to the cooling of the nebula and therefore also to the thermal equilibrium (see Sections 14.2.10 and 14.2.11). The chart in Figure 14.5 shows just the relevant small cut outs from the highly complex term diagrams [8], [31]. For the most

Table 14.3 Required ionization energy of some important elements and ionization stages compared to H and He.

Ion	[S II]	[N II]	[O III]	[Ne III]	[O II]	H II	He II	He III
E [eV]	10.4	14.5	35.1	41.0	13.6	13.6	24.6	54.4
λ [Å]	1193	855	353	302	911	911	504	227

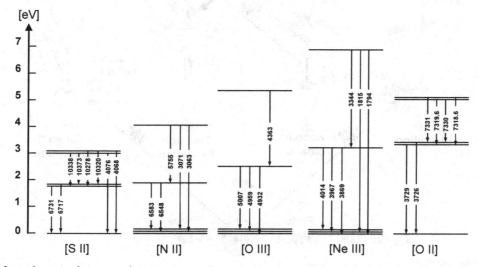

Figure 14.5 Cutouts from the term diagrams of important metal ions in emission nebulae with the required excitation energies in [eV] and the wavelengths of the forbidden transitions

important metal ions, the required excitation energies in [eV] and the wavelengths of the "forbidden" transitions are shown.

As a supplement Figure 14.6 presents a more detailed view on the transitions of the [O III] line. Due to the very low transition probability the 4932 Å line is omitted.

The three major transitions are observed in the optical wavelength domain and applied for temperature diagnostics, as explained later.

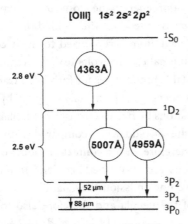

Figure 14.6 Forbidden emission lines of [O III]

1. Transitions between energy levels $^1S_0$ ➔ $^1D_2$ generating an emission line at 4363 Å.
2. Transitions between energy levels $^1D_2$ ➔ $^3P_1$ generating an emission line at 4959 Å.
3. Transitions between energy levels $^1D_2$ ➔ $^3P_2$ generating an emission line at 5007 Å.

Further, two minor transitions are temperature independent and used to measure the electron density. However, the according emissions appear in the far infrared range and are therefore not accessible by amateur means:

1. Transitions between energy levels $^3P_2$ ➔ $^3P_1$ generating an emission line at 52 μm.
2. Transitions between energy levels $^3P_1$ ➔ $^3P_0$, generating an emission line at 88 μm.

14.2.10 Scheme of the Fluorescence Processes in Emission Nebulae

In an emission nebula usually short-wavelength photons are transformed into less energetic ones with longer wavelength. This process is called fluorescence. The diagram in Figure 14.7

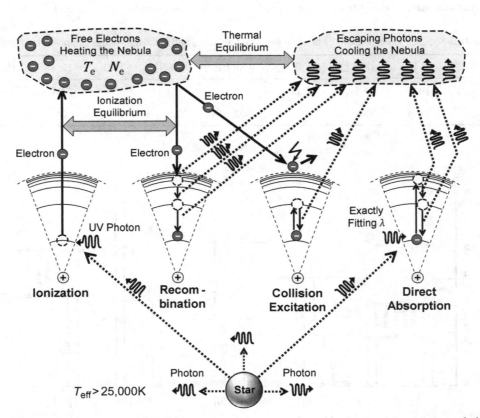

Figure 14.7 Schematic overview to the most significant fluorescence processes and equilibrium states in emission nebulae. Solid lines: moving electrons. Dotted lines: moving photons (R. Walker).

presents for an emission nebula the previously explained, individual processes in the entire context. Not shown here are bremsstrahlung processes, which become typically relevant mainly in supernova remnants (SNR) due to relativistic electron velocities. Processes with photons are shown here by dotted lines, those with electrons by solid lines. So-called bound-bound transitions between the electron shells are marked with black arrows. The broad, gray arrows show the two fundamentally important equilibrium states within the nebula.

The *thermal equilibrium* in the nebula between the heating process, generated by the kinetic energy of free electrons and the permanent removal of energy by escaping photons, regulates and determines the electron temperature T_e.

The *ionization equilibrium* between ionization and recombination regulates and determines the electron density N_e. If this balance is disturbed, the ionization zone of the nebula either expands or shrinks [8].

14.2.11 Cooling Mechanism by Forbidden Transitions

Here, with the example of the [O III] transitions, the important contribution of the forbidden transitions to the cooling mechanism in emission nebulae is considered more closely. The scheme in Figure 14.8 shows the nebular temperature

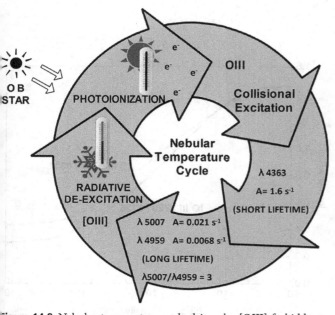

Figure 14.8 Nebular temperature cycle driven by [OIII] forbidden lines (M. Trypsteen). "A" is indicating here the transition probabilities in [s^{-1}].

cycle, displaying the photoionization of the hydrogen atoms by a neighboring OB type star, causing a fast rise of the electron temperature. The generated free electrons collide with already ionized metals such as O III (O^{++}). As mentioned earlier the kinetic energies of the free moving electrons are within the rough range of the differences 2.5 and 2.8 eV, indicated in Figure 14.6, causing collision-excitation up to the higher metastable energy levels $^1S_0$ and $^1D_2$. Depending on their electron population radiative de-excitation occurs which ends on one of the fine-structure levels of O III (O^{++}). This way observable emission lines are generated in the visible wavelength range. In the *Spectral Atlas* [1] most profiles of the numerous documented emission nebulae show the intense [O III] lines at λ5007 and λ4959, whereas λ4363 appears as a very tiny bump just in the highly excited planetary nebula NGC 7009. Therefore the emissions at λ5007 and λ4959 are here the main cooling forces. As a result of consecutive radiative emissions the electron temperature drops and an efficient "cooling" mechanism is activated. After several cycles a thermal equilibrium is reached and the electron temperature stabilizes (Figures 14.7 and 14.8).

14.2.12 Influences of T_e, N_e and Transition Probability A on the Cooling Mechanism

14.2.12.1 Influence of the Probability A

The probability of the individual electron transitions affects their rate which is expressed by the letter A with the unit "per second" [s^{-1}] and is further inversely proportional to the so called "lifetime," representing here the duration on a metastable level. Typical values for C-like $2sp^2$ and Si-like $3sp^2$ ions in the low probability range are 10^{-1}–10^{-12} s^{-1} and in the higher range 1–5 s^{-1} [19]. The higher transition probability of $^1S_0 \rightarrow {}^1D_2$ (λ4363) results in a shorter lifetime, in contrast to the transitions from $^1D_2$ down to the fine-structure levels $^3P_2$ and $^3P_1$, exhibiting relatively long lifetimes (λλ5007 and 4959). Therefore the chance for an untimely collision deexcitation on the $^1D_2$ energy level is significantly higher compared to the $^1S_0$ level.

The different transition probabilities of the [O III] emissions are already recognizable at a glance just considering the intensity ratios λ5007/λ4959 ≈ 3 which can be observed in all emission nebulae like M42 (*Spectral Atlas* [1]). This ratio corresponds here to the theoretically predicted probability ratio.

14.2.12.2 Influence of electron density N_e

As a consequence, at lower electron densities N_e which is typical for most emission nebulae, there will be more collision de-excitation for the $^1D_2$ energy level, compared to the $^1S_0$ level. Therefore less radiative de-excitation occurs here and the emissions $\lambda\lambda5007$ and 4959 weaken. Consequently collisional excitation from energy level $^1D_2$ up to $^1S_0$ also happens, which increases the population on the $^1S_0$ level and relatively intensifies the $\lambda4363$ line. Consequently with lower electron density N_e, the tiny diagnostic line $\lambda4363$ is increasing and becomes visible by amateur means but is limited to higher excited, hotter nebulae.

14.2.12.3 Influence of Electron Temperature T_e

As the temperature T_e increases the average electron velocity gets much higher. Therefore more energetic collisions occur with electrons on the $^1D_2$ level exhibiting relatively long lifetimes. As a result at low electron density, typical for planetary nebulae and H II regions, the weak $\lambda4363$ intensifies more, relative to $\lambda\lambda5007$ and 4959.

14.2.13 Estimation of T_e and N_e by the O III and N II Method

The known influences of T_e and N_e on the cooling mechanism can be used for estimations of electron temperature. Many methods have been described so far [8], [19]. However, due to the very weak diagnostic lines, these methods can be applied only to spectra with high resolution and quality, a real challenge with amateur equipment. Further, in the case of longer exposure times, the Hg I line $\lambda4358$, generated by light pollution, may interfere with the nearby diagnostic line $\lambda4363$. The following procedure is based on the fundamental equations by Gurzadyan [8], which applies for the O III method the lines at $\lambda\lambda$ 5007, 4959 and 4363, and for the N II method at $\lambda\lambda$ 6548, 6584 and 5755.

O III procedure:

$$\frac{I(5007)+I(4959)}{I(4363)} = 0.0753 \cdot \frac{1+2.67\cdot10^5\cdot\sqrt{T_e}/N_e}{1+2.3\cdot10^3\cdot\sqrt{T_e}/N_e}\cdot e^{33,000/T_e}$$

$$\{14.12\}$$

N II procedure :

$$\frac{I(6548)+I(6584)}{I(5755)} = 0.01625 \cdot \frac{1+1.94\cdot10^5\cdot\sqrt{T_e}/N_e}{1.03+3.2\cdot10^2\cdot\sqrt{T_e}/N_e}\cdot e^{25,000/T_e}$$

$$\{14.13\}$$

For the calculation of the electron temperature, these equations cannot be explicitly converted and solved by T_e and contain additionally the variable N_e. But for T_e, empirical equations exist which are valid for thin gases $N_e < 10^3\ cm^{-3}$ (typical for H II regions and SNR). ("ln" is the natural logarithm to base e.)

O III procedure :

$$T_e = \frac{33,000}{\ln\left(R_1/8.74\right)}$$

$$\{14.14\}$$

in which $R_1 = \dfrac{I(5007)+I(4959)}{I(4363)}$

N II procedure :

$$T_e = \frac{25,000}{\ln\left(R_2/9.85\right)}$$

$$\{14.15\}$$

in which $R_2 = \dfrac{I(6548)+I(6584)}{I(5755)}$

For the explicit calculation of N_e, with known T_e, here the Equations $\{14.12\}$ and $\{14.13\}$ are accordingly converted:

O III procedure :

$$N_e = \sqrt{T_e}\,\frac{2.67\times10^5 - \dfrac{3.05\times10^4 R_1}{(e^{33,000/T_e})}}{\dfrac{13.3\,R_1}{(e^{33,000/T_e})}-1}$$

$$\{14.16\}$$

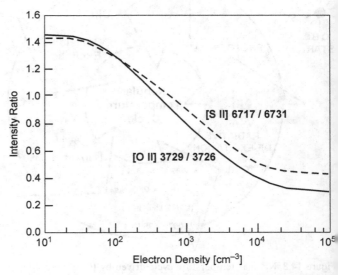

Figure 14.9 Estimation of the electron density N_e by the intensity ratio of the [S II] or [O II] emissions

N II procedure :
$$N_e = \sqrt{T_e}\,\frac{1.94 \times 10^5 - \dfrac{1.97 \times 10^4 R_2}{(e^{25,000/T_e})}}{\dfrac{61.5\,R_2}{(e^{25,000/T_e})} - 1}$$

$$\{14.17\}$$

If the recording of a spectral profile can be limited to a defined region within the nebula, in both Equations {14.16} and {14.17} the N_e variables become identical. Then N_e can be eliminated by equalization of Equations {14.16} and {14.17}. Anyway the implicitly remaining variable T_e requires solving the equation iteratively. However, this requires that the values of all diagnostic lines, for both methods, are available in good quality.

14.2.14 Estimation of the Electron Density by the S II and O II Ratio

The electron density N_e can be estimated with the diagram by Osterbrock in Figure 14.9 from the ratio of the two sulfur lines [S II] λλ 6716, 6731 or the oxygen lines [O II] λλ 3729, 3726. The big advantage of this method is that these lines are so close together that the extinction and instrumental responses cannot exert any significant effect on the ratio. The disadvantage is that both lines, except those generated by SNR, are generally very weak and therefore difficult to measure.

15 Amateurs and Astronomical Science

15.1 Participation in Astronomical Research

15.1.1 Astronomical Spectroscopy and the Pro–Am Culture

Precursor types of collaborations between amateur and professional astronomers date back to the eighteenth and nineteenth centuries. At that time individual and wealthy amateurs could afford the construction of an astronomical observatory equipped with the most advanced telescopes of that period and also among them, even employed professional astronomers. This way different forms of collaboration took place between amateurs, the academic world and other professional astronomical institutes, sometimes converging, other times diverging depending on the "hardware" capacities and personal points of view of all players [90].

Specific to the field of astronomical spectroscopy two English "pioneer" amateur astronomers can be referred. Sir William Huggins (1824–1910), a former businessman, was inspired by the spectroscopy tandem of that time, the German chemist Robert Bunsen (1811–1899) and the German physicist Gustav Kirchhoff (1824–1887). Besides taking spectra of the Sun, stars, nebulae and comets the most important innovation Huggins was responsible for occurred in 1868 – the accurate determination of the radial velocity of stars by measuring the Doppler shift of spectral lines. The same attempts were carried out at that time by the Italian astronomer Father Angelo Secchi (1818–1878) in 1866.

The second pioneer was Sir Joseph Norman Lockyer (1836–1920), formerly a clerk in the War Office in 1857, who discovered helium. His work was concentrated on solar spectroscopy and spectroscopic recordings to classify star spectra. At the start of the twentieth century the use of spectroscopy was booming and simultaneously there was a high demand for bigger apertures for telescopes. Even for the wealthiest amateur astronomers of that time it became too expensive to go further with this evolution. This caused a divergence between amateurs and professionals. Fortunately at that moment amateurs had the possibility to join astronomical associations for example the BAA or British Astronomical Association, founded in 1890, the AAVSO or American Association of Variable Star Observers, founded in 1911 [24]. This was necessary to offer the amateurs the possibility to upgrade their skill level as the astrophysical knowledge was increasing fast. Nevertheless, the astronomical spectroscopy was mostly practiced at the universities and astronomical institutes, where amateurs could contribute to the observations. An example is the call for amateurs organized by Pickering, director of the Harvard College observatory. In the second half of the twentieth century and with the start of the twenty-first century the technological evolution created a lot of new possibilities for amateurs. The large scale production of telescopes, the development of CCD cameras and the first affordable spectrographs for the amateur gave the impulse for a revival of the pro–am collaboration programs. The new technology made it possible for amateurs to deliver scientific results useful for professional astronomers.

Meanwhile, the great breakthrough of the pro–am collaboration idea in all socio-economic layers of society was created in the 1990s and astronomy was among the successful disciplines well suited for this type of collaboration [85]. Especially in Europe spectroscopic amateur groups made important progress in the field of pro–am collaborations on a high quality scientific level. This resulted in a continuously increasing number of scientific papers where amateurs contributed. From the interested and enthusiastic beginner, through to the dedicated and advanced amateur, more and more people will enter the level of contributing amateur working at almost a professional quality (Figure 15.1).

This work also makes part of the idea to improve the scientific skill level of the amateur interested in astronomical spectroscopy. As the number of potential "pro–am ready" amateurs can be increased also this way, it benefits the overall quality of the contribution to the scientific knowledge in astrophysics.

15.1.2 The Structure of a Pro–Am Collaboration

A good structure from the beginning is a prime requirement for the success of a pro–am collaboration. The diagram of

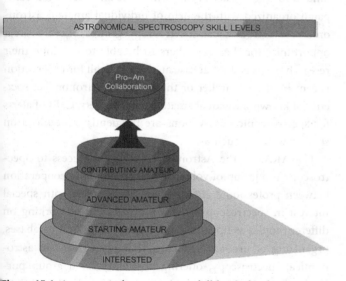

Figure 15.1 Astronomical spectroscopy skill levels for the amateur

Figure 15.2 introduces the different members of the pro–am collaboration such as the scientific staff, the astronomical observatory, the pro–am coordinator and the pro–am participants. Within this structure they work together following generally seven different phases.

Phase 1: Interested Amateurs

An important message to all interested amateurs that must be clear at the very beginning is that a pro–am collaboration is a high quality science project. The primary goal of this collaboration is to make a contribution to scientific work which can be a (funded) local or cross-border project. Sometimes it can also be organized as a contribution to a PhD or post-doctoral teamwork. The first task for the pro–am coordinator is to thoroughly inform interested amateurs about the philosophy of the pro–am collaboration to avoid misconceptions.

Phase 2: Amateur Team

The advanced and properly skilled amateurs form a team which is guided by the pro–am coordinator. This is also an ideal moment for the team members to share their knowledge and experiences in the field. Apart from the scientific "power" of the team it is simultaneously a good task for the pro–am coordinator to observe the social contacts between the different people in the team.

Phase 3: Project Presentation

In this third phase the pro–am project is presented in detail by the scientific staff. This staff is mostly composed of professional members working at universities and/or astronomical institutes. For the presentation of the project a meeting can be organized, although Internet based informative channels like a specific website of the project or a webinar on regular basis

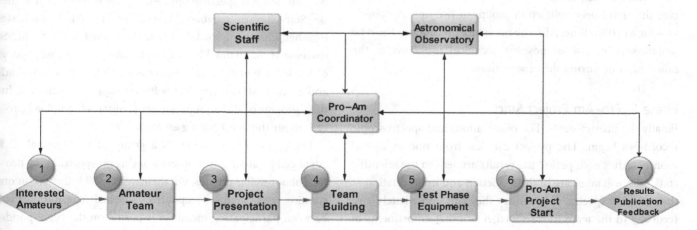

Figure 15.2 Diagram of the structure of a pro–am collaboration

are good alternatives. The main goal here is to inform the team members in as much detail as possible on what are the project's aims, the procedures to follow in practice, the manuals of the scientific instruments (telescope guiding, spectrographs and cameras) and the practical organization of the observation times. Different methods of organizing a pro–am project are possible depending on the location of the observation. This is mostly an astronomical observatory or institute where the team works together to make the observations and recordings. A second method is individual participation to the project with contributors using personal equipment at different locations. Here, the use of equivalent types of telescope, spectrograph and camera is recommended. A third method and recent evolution is the remote way of participation. Here, fixed astronomical equipment and/or pier is hired on a perfect location suitable for astronomical observations. Access for the team members to the telescope, spectrograph and camera is controlled by an Internet connection.

Phase 4: Team Building

This phase is facultative depending on the location of the observations. The pro–am coordinator normally makes contact with the astronomical observatory where the observations will be carried out. This can be an astronomical institute funded by one or more countries. If it is possible it is certainly recommended to visit the location of observation to be familiarized with the equipment and the way of working of the observatory or institute.

Phase 5: Test Phase Equipment

Now this is the time to prepare for the real work. The team makes a few recordings (photography, spectra) to test the equipment. The final time schedule for the observations must be discussed so the teams will be ready to relieve each other in case they have been split up in groups of two or three observers for a certain time. Also the necessary precautions must be undertaken in case of possible technical problems at the moment of or during the observations.

Phase 6: Pro–Am Project Start

Finally the project starts. The observations and spectroscopic recordings begin. The project can last from one to several months. After each period the results are sent to the scientific staff, who will take care of the reduction and interpretation of the recorded spectra. If necessary they can immediately give feedback to the team in case further or other adjustments of the equipment and the recordings are needed.

Phase 7: Results, Publication, Feedback

Afterwards the results of the project are published as a scientific poster at a symposium or seminar and/or as a paper in a scientific journal. A good idea is to ask all team members to fill in a survey on the pros and cons they have experienced. This feedback can be used later to ameliorate the quality of future pro–am projects.

15.2 Observation Campaigns for Amateurs

In astronomical research programs it is usually the task of the researcher or their team to make a reservation for observation time at a telescope institute. The high demand for observation times worldwide forces the telescope institutes to start ranking the requests, so a "scientific queue" is created. Another limiting factor is the budget, which is sometimes limited for the research team. Therefore an observation campaign organized with the help of individual amateur astronomers or astronomical spectroscopy groups offers a great opportunity for the researchers to be able to continue their research projects. In that case an alert or a call for observation is sent by the researcher or their team to astronomical societies of known individual amateur astronomers. Calls of alerts for spectroscopic observations are frequently announced on specific websites such as:

The ARAS or the Astronomical Ring for Access to Spectroscopy is a group of volunteers to promote the cooperation between professional and amateur astronomers with special interest in spectroscopy. It has a broad field of working on different topics as updating and managing spectral databases, organizing of pro–am seminars, training sessions in astronomical spectroscopy, offering possibility for a group purchase of spectrographs, offering a spectral upload facility to contributing amateurs and connectivity via a global network on astronomical spectroscopy. An excellent realization is the Be Star Spectra database, known as the BESS database together with the upload facility tool ARASBEAM. The BESS database is maintained by the LESIA team at the observatory of Meudon, France. The protocol is nowadays widely spread and even most astronomical software suppliers include it in their programs, which impressively facilitates the upload procedure for the contributing amateur [86].

The CONVENTO group is a group of professional and advanced amateur astronomers working on spectroscopy projects about stellar physics. On their website all information on running campaigns is updated. Internal communication between participating members is posted on the corresponding discussion forum [87].

The AAVSO also has an "observing campaign" link on their website, where researchers can post a call for spectroscopic surveys [24].

15.3 Contributions by Amateurs

15.3.1 Short-term Campaign (Months)

Topic: Spectroscopic observing campaign of WR140 during its periastron passage

WR140 is a hot, luminous colliding wind binary, which is composed of a Wolf Rayet WC7pd star and an O5 main sequence star (see [1] Plate 33). The orbit eccentricity is 0.88 which shows strongly elliptic movements. The distance between the two stars varies from 30 AU at the apastron to only 2 AU at the periastron. With a periodicity of 7.9 years the periastron passage is always a highlight for astronomical spectroscopy. The periastron passage series of 1993, 2001 and 2009 revealed important information about the orbital parameters, refined values of both stars' masses and for both stars the corresponding mass-loss rates. The periastron passage moment is also accompanied by an impressive dust formation, causing an infrared excess measured in the spectrum. This is caused by back radiation of the colliding wind driven condensed dust carbon particles which absorbed radiation from the star. To better understand the physical properties of the star system and to refine the astrophysical model, further spectroscopic data are necessary, which can be collected from observations during the future periastron passages of 2017, 2025, 2032. These passages are great

opportunities to organize pro–am collaborations, based on the experiences of the successful campaign of 2009, also known as the MONS campaign [88].

15.3.2 Long-term Campaign (Years)

Topic: Be stars observing campaign

Some spectral type B stars show a specific behavior. At a certain stage they start to emit radiation and are called Be stars [1]. This is caused by the formation of a circumstellar decretion disk. The combination of the rapid rotation, generated magnetic fields and radial pulsations contribute to the ejection of material from the star. A typical example is Gamma Cassiopeiae. See *Spectral Atlas*, Plate 31. Depending on the equatorial thickness of the circumstellar disk the Be star can additionally show absorption lines which is called the Be-shell stage. The transitions between the different phases, B–Be–Be-shell, are not completely understood. Therefore spectral surveys are needed. In 2006, a French group of amateur astronomers started spectral recordings with Lhires III spectrograph. In 2007 the BESS database was created and contains actually almost 50,000 recorded spectra of Be stars. Besides the Be star surveys another pro–am project concentrates on discoveries of new Be stars [20].

Such successful pro–am collaborations show the possibilities of this new form of "citizen science" model guided by the modern technological evolution [89]. The more amateurs entering the group of contributing spectroscopists worldwide, the more successful will be the pro–am project.

APPENDIX A:
Abbreviations, Acronyms and Common Units

This list contains abbreviations, acronyms and a selection of common units and important variables

A	probability of an electron transition [s^{-1}]
Å	unit of wavelength: ångstrom
AAVSO	American Association of Variable Star Observers
ADC	analog-to-digital converter of an image sensor
ADU	analog/digital units, intensity unit for unprocessed raw profiles
AGB	asymptotic giant branch (HRD)
AGN	active galactic nuclei
ARAS	Astronomical Ring for Access to Spectroscopy
ASCOM	open source software platform for telescope guiding
AU	astronomical unit, 149.6×10^6 km
B	magnetic flux density
BASS	Basic Astronomical Spectroscopy Software
BeSS	Be Star Spectra database (laboratoire LESIA de l'Observatoire de Paris-Meudon)
BJ	Balmer jump
c	speed of light 300,000 km s^{-1}
CCD	Charge-coupled device, sensor for digital imaging
CDS	Centre de Données astronomiques de Strasbourg
CFL	compact fluorescent lamps
cgs	system of units: centimeter, gram, second
D	angular dispersion
D	Balmer decrement
D	distance
D	effective aperture or diameter of a telescope's objective
d	days
d	distance between adjacent grooves of a grating, reciprocal of the grating constant
DIY	do it yourself
DSLR	digital single-lens reflex (camera)
erg	cgs unit for energy

ES	electron spin
ESL	energy saving lamps
ESO	European Southern Observatory
ETC	exposure time calculator (SimSpec)
EW	equivalent width (of a spectral line) [Å]
f	focal length of an optical system
F	flux
FOV	field of view
FSR	free spectral range (without overlapping)
FWHM	full width (of a spectral line) at half maximum height [Å]
FWZI	full width (of a spectral line) at zero intensity [Å]
G	gravitational constant
GC	international glass code
GCA	specification of refractive index, glass code A
GCB	specification of the Abbe number V_d, glass code B
GTR	general theory of relativity
Gy	gigayear (1 billion y)
h	hours
H$_{(0)}$	Hubble constant or present value of the Hubble parameter
H$_{(t)}$	Hubble parameter depending on a certain epoch
HB	horizontal branch (HRD)
HRD	Hertzsprung–Russell diagram
HWHD	half width (of a spectral line) at half depth (equivalent to HWHM)
	HWHM half width (of a spectral line) at half maximum height (equivalent to HWHD)
HWZI	half width (of a spectral line) at zero intensity [Å]
I	intensity
IAU	International Astronomical Union
Ir	correction function

IR	infrared range of the electromagnetic spectrum	P	reciprocal linear dispersion
IRAF	software for the reduction and analysis of astronomical data (by NOAO)	PA	position angle (galaxy, binary system)
		pc	parsec, 1 pc $\approx$ 3.26 ly, Mpc = megaparsec
IRIS	astronomical images processing software by Christian Buil	PHD	autoguiding software
		PMS	pre-main sequence star, young protostar, not yet established on main sequence
ISIS	spectroscopic analysis software by Christian Buil		
		PN	planetary nebula
ISM	interstellar matter	QE	wavelength dependent quantum efficiency of an image sensor
J	joule, SI unit of energy		
K	kelvin, temperature unit K $\approx$ °Celsius + 273	R	resolving power of a spectrograph
L	luminosity	R_i	residual intensity
LBV	luminous blue variable	R_s	Strömgren radius
LCAO	linear combination of atomic orbitals	RGB	red giant branch (HRD)
LD	line depth (of an absorption line)	s	seconds
LDR	line depth ratio (LD in relation to the continuum intensity)	SB	surface brightness (of 2D appearing objects)
		SB1	SB1 system. Spectroscopic binary stars with a strong brightness difference. Just the spectrum of the brighter component can be observed.
ly	light year, 1 ly $\approx$ 9.46 × 10^{12} km		
M	million (mega)		
M	mass	SB2	SB2 system. Spectroscopic binary stars with a small brightness difference. A composite spectrum of both components is observable.
M	absolute magnitude, apparent brightness at a distance of 10 parsec [mag]		
M_B	absolute magnitude in the blue range of the spectrum [mag]	SE	Schrödinger equation
		SN	supernova explosion
M_i	initial stellar mass	SNR	supernova remnant
M_V	absolute magnitude in the visual (green) range [mag]	SNR	signal-to-noise ratio (level of a desired signal to the level of background noise)
m	order of diffraction	STR	special theory of relativity
m_B	apparent magnitude in the blue range [mag]	T_e	Temperature of an electron gas [K]
m_V	apparent magnitude in the visual (green) range [mag]	T_{eff}	effective temperature of a stellar photosphere [1]
MIDAS	ESO-MIDAS, software for reduction and analysis of astronomical data	TDSE	time dependent form of the Schrödinger equation (SE)
MO	molecular orbit method	TISE	time independent form of the Schrödinger equation (SE)
MS	main sequence, position of stars in the HRD with luminosity class V		
		UHF	ultra high frequency
N	focal ratio, whereas N = f/D, N is also expressed as f-number "f/N"	UV	ultraviolet range of the electromagnetic spectrum
N	total number of grooves (grating)	v_r	spectroscopically measured radial velocity
N_e	Density of an electron gas [cm^{-3}]	v_∞	terminal velocity of a stellar wind or an expanding envelope
n_{index}	refractive index of a certain medium (for air $n \approx 1$)		
		VPHG	volume-phase holographic gratings
NED	NASA Extragalactic Database	Vspec	spectroscopic analysis software "Visual Spec" by Valérie Desnoux
nm	nanometer		
NS	spin of the nucleus		
Ω	cosmological density parameter "omega"	WMAP	NASA Wilkinson Microwave Anisotropy Probe Mission
Or	original stellar spectrum		
P	peak intensity	WMO	World Meteorological Organization

WR	Wolf–Rayet stars
y	year
Z	metal abundance (metallicity) [1]
z	z-value, a measure for the redshift, distance and the past
ZAMS	zero age at main sequence – young star just joining the main sequence

$\odot$	comparison to the Sun: solar mass $M_\odot$, luminosity $L_\odot$ and radius $R_\odot$
2D	two dimensionally
3D	three dimensionally
200 L, 900 L	Reflection gratings with 200 or 900 lines mm^{-1}

APPENDIX B:
Absolute Magnitudes of Main Sequence Stars

This table shows the approximate absolute magnitudes (M) for the spectral classes of the main sequence stars [7]. Supplemented are some literature values for some giants (III) and supergiants (I), exhibiting a considerable scattering range.

Spectral Class	Main Sequence (V) [M]	Giants (III) [M]	Supergiants (I) [M]
O5	−5.5		
O6	−5.3		
O7	−4.8	68 Cygni −6.7	
O8	−4.4	Meissa, λ Ori −4.3	
O9	−4.3	Iota Ori −5.3	Alnitak, ζ Ori −5.3
B0	−4.1	Alnilam, ε Ori −6.7	
B1	−3.5	Alfirk, β Cep −3.5	
B2	−2.5	Bellatrix, γ Ori −2.8	
B3	−1.7		
B5	−1.1	δ Per −3.0	Aludra, η CMa −7.5
B6	−0.9		
B7	−0.4	Alcyone, η Tau −2.5	
B8	0.0	Atlas, 27 Tau −2.0	Rigel, β Ori −6.7
B9	0.7		
A0	1.4		
A1	1.6		
A2	1.9		Deneb, α Cyg −8.7
A3	2.0		
A5	2.1	α Oph 1.2	
A7	2.3	γ Boo 1.0	
A9	2.5		
F0	2.6	Adhafera, ζ Leo −1.0	
F1	2.8		
F2	3.0	Caph, β Cas 1.2	
F3	3.1		
F4	3.3		

(*cont.*)

Spectral Class	Main Sequence (V) [*M*]	Giants (III) [*M*]	Supergiants (I) [*M*]
F5	3.4		Mirfak, α Per −4.5
F6	3.7		
F7	3.8		
F8	4.0		Wezen, δ CMa −6.9
F9	4.2		
G0	4.4		Sadalsuud, β Aqr −3.3
G1	4.5		
G2	4.7		Sadalmelik, α Aqr −3.9
G5	5.2		
G6	5.3		
G7	5.5	Kornephoros, β Her −0.5	
G8	5.6	Vindemiatrix, ε Vir 0.4	
G9	5.7		
K0	5.9	Dubhe, α Uma −1.1	
K1	6.1	Arcturus, α Boo −0.3	
K2	6.3	Cebalrai, β Oph 0.8	
K3	6.9		
K4	7.4		
K5	8.0	Aldebaran, α Tau −0.7	
K7	8.5	Alsciaukat, α Lyn −1.1	
M0	9.2		
M1	9.7	Scheat, β Peg −1.5	Antares, α Sco −5.3
M2	10.6		Betelgeuse, α Ori −5.3
M3	11.6		
M4	12.9		
M5	14.5		Ras Algethi, α Her −2.3
M6	16.1		

APPENDIX C:
The Solar Echelle Spectrum: An Aid to Orientation

Recorded by SQUES echelle spectrograph, individual orders have been processed by IRIS.

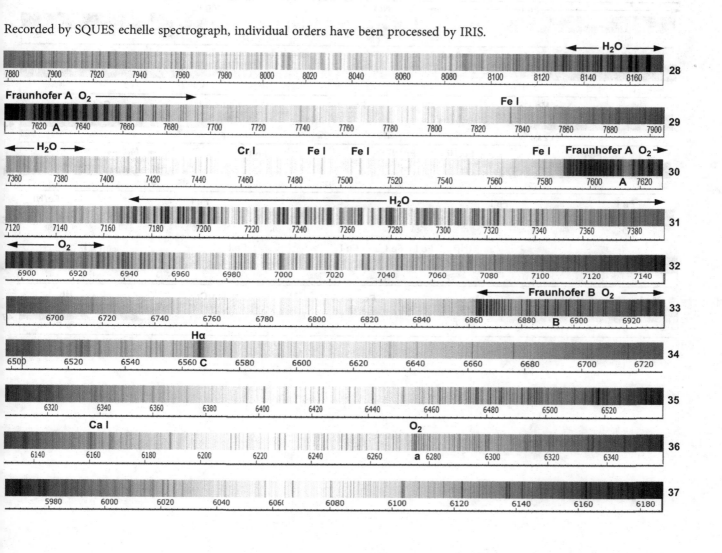

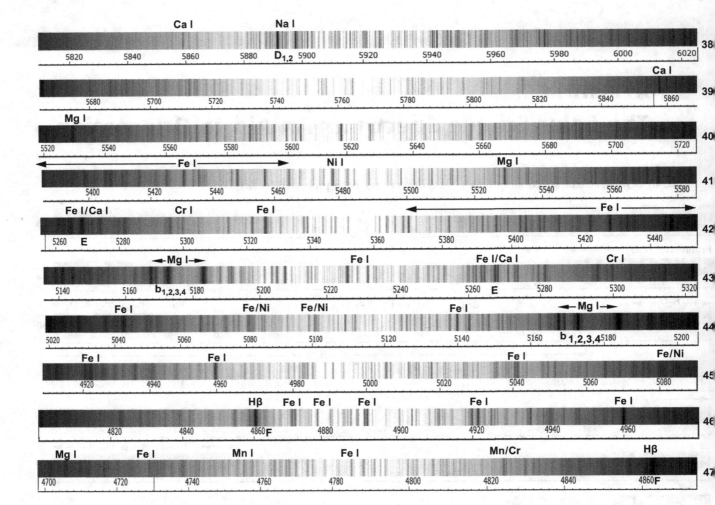

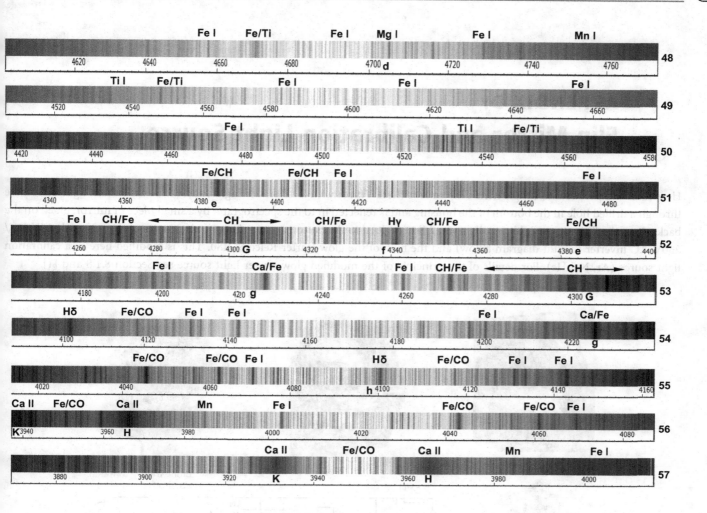

APPENDIX D:
Flip Mirror and Calibration Light Source

Here follows a proposal to supplement a flip mirror (here Vixen) with a calibration light source. The calibration light is fed here through a drilled hole in the bottom of the housing and is reflected in to the spectrograph by a small piece of mirror, glued on the backside of the flip mirror. Additionally a small box is glued to the bottom, containing the electronics of the self made 12V DC/ 230V AC inverter (circuit diagram below) and the bulb of the glow starter Relco SC480. This is modified here as a calibration light source (Ar, Ne, He). For details of the principle of the modified glow starter light source see Section 8.1.8 and [1].

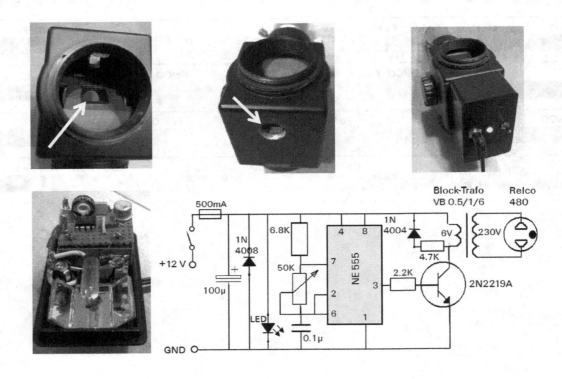

REFERENCES AND FURTHER READING

Literature

[1] R. Walker, *Spectral Atlas for Amateur Astronomers: A Guide to the Spectra of Astronomical Objects and Terrestrial Light Sources* (Cambridge: Cambridge University Press, 2017). This atlas is supported by the present book to provide basic information, definitions and formulas.

[2] O. Struve, V. Zeberg, *Astronomy of the 20th Century* (New York: Macmillan, 1962).

[3] M. Asplund *et al.*, The Chemical Composition of the Sun (2009), available online at: https://arxiv.org/pdf/0909.0948.pdf

[4] I. Appenzeller, *Introduction to Astronomical Spectroscopy* (Cambridge: Cambridge University Press, 2012).

[5] T. Eversberg, K. Vollmann, *Spectroscopic Instrumentation, Fundamentals and Guidelines for Astronomers* (Berlin, Heidelberg: Springer, 2015).

[6] D. F. Gray, *The Observation and Analysis of Stellar Photospheres*, Third Edition (Cambridge: Cambridge University Press, 2008).

[7] R. O. Gray, Ch. J. Corbally, *Stellar Spectral Classification* (Princeton: Princeton University Press, 2009).

[8] G. A. Gurzadyan, *The Physics and Dynamics of Planetary Nebulae* (Springer, 2010).

[9] J. B. Hearnshaw, *The Analysis of Starlight: One Hundred and Fifty Years of Astronomical Spectroscopy* (Cambridge: Cambridge University Press, 1990, 2014).

[10] J. B. Hearnshaw, *Astronomical Spectrographs and Their History* (Cambridge: Cambridge University Press, 2009).

[11] J. B. Kaler, *Stars and Their Spectra, An Introduction to the Spectral Sequence* (Cambridge: Cambridge University Press, 2011).

[12] A. K. Pradhan, S. N. Nahar, *Atomic Astrophysics and Spectroscopy* (Cambridge: Cambridge University Press, 2011).

[13] K. Robinson, *Spectroscopy, The Key to the Stars* (New York: Springer, 2007).

[14] J. P. Rozelot, C. Neiner, *The Rotation of Sun and Stars, Lecture Notes in Physics 765* (Berlin Heidelberg: Springer, 2009).

[15] S. F. Tonkin, *Practical Amateur Spectroscopy* (London: Springer, 2004).

[16] S. Svanberg, *Atomic and Molecular Spectroscopy, Basic Aspects and Practical Applications* (Berlin Heidelberg: Springer, 2012).

[17] J. Tennyson, *Astronomical Spectroscopy, 2nd Edition* (Singapore: World Scientific Publishing Co., 2011).

[18] W. Hergert, T. Wriedt, *The Mie Theory, Basics and Applications* (Berlin Heidelberg: Springer Verlag, 2012). (Mieplot)

[19] D. E. Osterbrock, G. J. Ferland, *Astrophysics of Gaseous Nebulae and Active Galactic Nuclei*, Second Edition (Mill Valley, CA: University Science Books, 2005).

[20] C. Neiner *et al.*, *Active OB Stars: Structure, Evolution, Mass-Loss and Critical Limits*, IAU Symposium 272 (Cambridge: Cambridge University Press, 2011).

[21] E. Niemczura *et al.*, *Determination of Atmospheric Parameters of B-, A-, F- and G-Type Stars* (Springer International, 2014).

[22] T. Kogure, K.-C. Leung, *The Astrophysics of Emission-Line Stars*, (New York: Springer-Verlag, 2007).

Internet Sources

The following publications and presentations are from various internet sources. The web addresses may change or expire with time. The corresponding PDF or html files should readily be found using a search engine, by entering of title and author.

Databases and Catalogs

[23] A. Lobel, *SpectroWeb, The Interactive Database of Spectral Standard Star Atlases*, Royal Observatory of Belgium. http://alobel.freeshell.org/ http://spectra.freeshell.org/SpectroWeb_news.html

[24] AAVSO, American Association of Variable Star Observers https://www.aavso.org/

[25] CDS Strasbourg: *SIMBAD Astronomical Database* http://simbad.u-strasbg.fr/simbad/

[26] NASA Extragalactic Database (NED) http://nedwww.ipac.caltech.edu/

[27] NIST Atomic Spectra Database: http://physics.nist.gov/PhysRefData/ASD/lines_form.html

Stellar Magnetic Fields

[28] E. Landi Degl'Innocenti, On The Effective Landé Factor of Magnetic Lines, *Solar Physics*, 77, 1–2, (1980), 285–9, available online at: http://adsabs.harvard.edu/full/1982SoPh...77..285L

[29] G. Mathys *et al.*, The Mean Magnetic Field Modulus of Ap Stars, *Astronomy and Astrophysics*, 123, 2, (1996), 353–402, available online at: http://aas.aanda.org/articles/aas/abs/1997/08/ds1257/ds1257.html

[30] A. Reiners, Magnetic Fields in Low-Mass Stars: An Overview of Observational Biases (2014), available online at: http://arxiv.org/pdf/1310.5820v1.pdf

Emission Nebulae

[31] F. Gieseking *Planetarische Nebel*, six-part article series, in "Sterne und Weltraum," 1983

[32] J. Schmoll, AIP: 3D Spektrofotometrie Extragalaktischer Emissionslinienobjekte, PhD Thesis, University of Potsdam (2001), available online at: http://www.aip.de/groups/publications/schmoll.pdf

Processing Calibration and Normalization of Spectral Profiles

[33] G. Aldering, *SN Factory Spectrophotometry Requirements Document* (2000), available online at: http://snfactory.lbl.gov/snf/ps/flux_calib.ps

[34] ESO/UVES: *Uves Quality Control: Flux Calibration, Response Curves*, available online at: http://www.eso.org/observing/dfo/quality/UVES/pipeline/response.html

[35] W. Stubbs *et al.*, Addressing the Photometric Calibration Challenge: Explicit Determination of the Instrumental Response and Atmospheric Response Functions, and Tying it All Together (2012), eprint arXiv:1206.6695, available online at: http://arxiv.org/abs/1206.6695

[36] W. Stubbs *et al.*, Toward 1% Photometry: End-to-end Calibration of Astronomical Telescopes and Detectors (2006), available online at: http://arxiv.org/pdf/astro-ph/0604285v1.pdf

[37] J. Valenti, SISD Training Lectures in Spectroscopy: Anatomy of a Spectrum (STSCI) available online at: http://www.stsci.edu/hst/training/events/Spectroscopy/Spec02Nov09.pdf

Radial Velocity

[38] C. Chubak *et al.*, Precise Radial Velocities of 2046 Nearby FGKM Stars and 131 Standards (2012), available online at: http://arxiv.org/pdf/1207.6212v2.pdf

Rotational Velocity

[39] F. Fekel, Rotational Velocities of B, A, and Early-F Narrow-lined Stars (2003), available online at: http://www.jstor.org/stable/10.1086/376393

[40] F. Fekel, Rotational Velocities of Late Type Stars (1997), available online at: http://articles.adsabs.harvard.edu/full/1997PASP..109..514F

[41] R. W. Hanuschik, Stellar V sin i and Optical Emission Line Widths in Be Stars (1989), available online at: http://articles.adsabs.harvard.edu/full/1989Ap%26SS.161...61H

[42] N. A. Moskovitz *et al.*, Characterizing the Rotational Evolution of Low Mass Stars: Implications for the Li-rich K-giants (2009), available online at: http://eo.nso.edu/sites/eo.nso.edu/files/files/ires/IRES_2007/Moskovitz_TechRpt.pdf

[43] A. Slettebak, Determination of Stellar Rotational Velocities (1985), available online at: http://articles.adsabs.harvard.edu/full/1985IAUS..111..163S/0000164.000.html

[44] M. Trypsteen, Determination of Planetary Rotations with BASS, Vspec and ESO-MIDAS, *Guidestar Magazine*, 9, (2015), 22–9, available online at: https://issuu.com/guidestar/docs

P Cygni Profiles

[45] M. Hogerheijde, *Radiative Processes*, Appendix E, Extra Problem Set, (Leiden Observatory), available online at: http://home.strw.leidenuniv.nl/~michiel/ismclass_files/radproc07/extra_problem_set_3.pdf

Measurement of Temperature

[46] K. Biazzo *et al.*, Effective Temperature vs Line-Depth Ratio for ELODIE Spectra, Gravity and Rotational Velocity Effects (2007), available online at: http://arxiv.org/pdf/0704.1456v1.pdf

[47] S. Catalano, *et al.*, Measuring Starspot Temperature from Line Depth Ratios, Part I (2002), available online at: http://www.aanda.org/articles/aa/pdf/2002/42/aa2543.pdf

[48] S. Catalano *et al.*, Measuring Starspot Temperature from Line Depth Ratios, Part II (2004), available online at: http://www.aanda.org/articles/aa/pdf/2005/11/aa1373.pdf

Spectroscopic Binary Systems

[49] H.-G. Reimann, *Kompendium für das Astronomische Praktikum*, Spektroskopische Doppelsterne, Lecture University Jena (in German), available online at: http://www.astro.uni-jena.de/Teaching/Praktikum/pra2002/node156.html

Balmer Decrement and Interstellar Extinction

[50] A. Acker *Spectrometry of Nebulae*, University of Strasbourg (2011), available online at: http://gawws.u-strasbg.fr/Stuff/Acker.pdf

[51] F. Arenou *et al.*, A Three-Dimensional Galactic Extinction Model, *Astronomy and Astrophysics*, 258, 1, (1992), 104–11, available online at: http://articles.adsabs.harvard.edu/full/1992A%26A...258..104A

[52] M. Brocklehurst, Calculations of Level Populations for the Low Levels of Hydrogenic Ions in Gaseous Nebulae, *Monthly Notices of the Royal Astronomical Society*, 153, (1971), 471–90, available online at: http://adsabs.harvard.edu/full/1971MNRAS.153..471B

[53] C. S. Reynolds *et al.*, A Multiwavelength Study of the Seyfert 1 Galaxy MCG-6-30, *Monthly Notices of the Royal Astronomical Society*, 2911, (1997), 403, available online at: http://adsabs.harvard.edu/abs/1997MNRAS.291..403R

Fluorescence Processes Emission Nebulae

[54] H. P. Gail, *Konstruktion eines einfachen Modellprogramms für die Struktur eines Gasnebels* (2006), University of Heidelberg (in German), available online at: www.ita.uni-heidelberg.de/~gail/hiiprog.ps

Spectrographic Measurement of Distances by Absolute Magnitudes

[55] G. Pace *et al.*, The Wilson–Bappu Effect: A Tool to Determine Stellar Distances, *Astronomy and Astrophysics*, 401, (2003), 997–1007, available online at: http://www.aanda.org/articles/aa/pdf/2003/15/aah3836.pdf

[56] G. A. Walker, Ch. G. Millward, A Convincing Mv-W(Hγ) Calibration for A and B Supergiants, *Astrophysical Journal*, 289, (1985), 669–75, available online at: http://adsabs.harvard.edu/full/1985ApJ...289..669W

Spectrographic Measurement of Surface Gravity

[57] S. Park *et al.*, Wilson–Bappu Effect: Extended to Surface Gravity (2013), available online at: http://arxiv.org/pdf/1307.0592v1.pdf

Spectrographs

[58] DADOS and BACHES Spectrographs, Baader Planetarium: http://www.baader-planetarium.de

[59] Shelyak Instruments: ALPY, Lhires III, LISA and eShel http://www.shelyak.com/

[60] SQUES Echelle Spectrograph, Eagleowloptics Switzerland http://www.eagleowloptics.com/

[61] BESOS Spectrograph, CAOS Group http://www2.astro.puc.cl/ObsUC/images/d/da/BESOS_a_prism_spectrograph_.pdf

[62] Rainbow Optics Star Spectroscope, www.starspectroscope.com

[63] Star Analyser, Paton Hawksley Education Ltd (PHEL) www.patonhawksley.co.uk

[64] RS Spectroscope, Rigel System https://www.optcorp.com/rigel-systems-rs-spectroscope.html

[65] Minispec, Astro Spectroscopy Instruments EU, http://www.astro-spec.com/de

[66] Spectra L200, JTW Astronomy http://www.jtwastronomy.com/products/spectroscopymain.html

[67] Trypsteen M., The Spectra-L200, part 1: The Ultimate Spectrograph, *Guidestar Magazine*, 9, (2014), 76–81; The Spectra-L200, part 2: In Practice, *Guidestar Magazine*, 10, (2014), 40–5, available online at: https://issuu.com/guidestar/docs

[68] Starlight Xpress SX, Starlight Xpress Ltd, UK, www.sxccd.com

Spectrographic Software

[69] MIDAS, ESO http://www.eso.org/sci/software/esomidas//

[70] IRAF, NOAO, http://iraf.noao.edu

[71] IRIS and ISIS: Webpage of Christian Buil http://www.astrosurf.com/buil/

[72] RSpec: Webpage of Tom Field http://www.rspec-astro.com/

[73] SpectroTools: Freeware Program by P. Schlatter http://www.peterschlatter.ch/SpectroTools/

[74] Visual Spec "Vspec": Webpage of Valerie Désnoux http://astrosurf.com/vdesnoux/

[75] SimSpec, Spectrograph Exposure Time Calculators (ETC) by Christian Buil http://www.astrosurf.com/buil/us/compute/compute.htm

[76] BASS, Basic Astronomical Spectroscopy Software, http://www.aesesas.com/mediapool/142/1423849/data/DOCUMENTOS/BASS_Project_1_.pdf

[77] M. Trypsteen, Astrospectroscopy@Cyberspace, *Guidestar Magazine*, 1, (2015), 80–4, available online at: https://issuu.com/guidestar/docs

Gratings

[78] Edmund Optics Inc http://www.edmundoptics.com/optics/

[79] Richardson Gratings, Newport Corporation http://www.gratinglab.com/Home.aspx

[80] Thorlabs Inc https://www.thorlabs.com/

Optical Design of Spectrographs

[81] D. J. Schroeder, R. L. Hilliard, Echelle efficiencies: Theory and Experiment, *Applied Optics*, (1980), available online at: https://www.osapublishing.org/ao/abstract.cfm?uri=ao-19-16-2833

[82] V. Stanishev, Correcting Second-Order Contamination in Low-Resolution Spectra (2007), available online at: http://arxiv.org/pdf/0705.3441v1.pdf

[83] S. Barnes, *The Design and Performance of High Resolution Echelle Spectrographs in Astronomy*, (University of Canterbury, NZ, 2004), available online at: http://ir.canterbury.ac.nz/handle/10092/5576

[84] L. Cowley, *Atmospheric Optics* website available online at: http://www.atoptics.co.uk/

Pro–Am Collaboration

[85] Ch. Leadbeater, P. Miller, The Pro–Am Revolution, *Demos* (2004), available online at: http://libweb.surrey.ac.uk/library/skills/Public%20understanding%20of%20science/promo_1.pdf

[86] ARAS, Astronomical Ring for Access to Spectroscopy http://www.astrosurf.com/aras/

[87] The Convento Group http://www.stsci.de/convento/

[88] R. Fahed *et al.*, The WR 140 Periastron Passage 2009: First Results from MONS and Other Optical Sources, eprint arXiv:1101.1430, available online at: http://arxiv.org/abs/1101.1430

[89] J. Hattenbach, Deciphering Starlight, *Sky & Telescope*, Sept., (2013), available online at: http://de.scribd.com/doc/162786805/Sky-Telescope-September-2013-Gnv64#scribd

[90] D. Boyd, Pro-Am Collaboration in Astronomy: Past, Present and Future, *Journal of the British Astronomical Association*, 121, 2, (2011), 73-90, available online at http://adsabs.harvard.edu/full/2011JBAA..121...73B

[91] J. R. Torres, On the Prediction of Visibility for Deep-Sky Objects, *Pleiades*, 1, 1, (2000), 2, available online at: http://www.uv.es/jrtorres/visib.pdf

[92] J. Dachs, R. Hanuschik *et al.*, Geometry of Rotating Envelopes around Be Stars Derived from Comparative Analysis of H-alpha Emission Line Profiles, *Astronomy and Astrophysics*, 159, 1–2, (1986), 276–90, available online at: http://articles.adsabs.harvard.edu//full/1986A%26A...159..276D/0000280.000.html

[93] S. Stefl *et al.*, V/R Variations of Binary Be Stars, *Active OB-Stars: Laboratories for Stellar and Circumstellar Physics*, ASP Conference Series, 361, (2007), 274, available online at: http://adsabs.harvard.edu/full/2007ASPC..361..274S

[94] S. S. Huang, Profiles of Emission Lines in Be Stars, *Astrophysical Journal*, 171, (1972), 549, available online at: http://cdsads.u-strasbg.fr/abs/1972ApJ...171..549H

Further Reading

[95] K. M. Harrison, *Astronomical Spectroscopy for Amateurs* (New York: Springer, 2011).

[96] C. R. Kitchin, *Optical Astronomical Spectroscopy* (New York: Taylor and Francis Group, 1995).

[97] J. P. Rozelot, C. Neiner, *Astronomical Spectrography for Amateurs*, EAS Publication Series, Volume 47 (2011).

[98] W. C. Seitter, *Atlas für Objektiv Prismen Spektren: Bonner Spektralatlas* (Astronomical Institute Bonn, 1975), out of print but available online at: http://www.archive.org/search.php?query=bonner%20atlas

[99] B. D. Warner, *A Practical Guide to Lightcurve Photometry and Analysis* (New York: Springer, 2006).

[100] H. Geiger, K. Scheel, *Geschichte der Physik Vorlesungstechnik*, (Springer, 1926).

[101] J. E. Dyson, D. A. Williams, *The Physics of the Interstellar Medium* (Manchester: Manchester University Press, 1980).

[102] J. C. Morrison, *Modern Physics for Scientists and Engineers*, second edition (Elsevier Academic Press, 2015).

[103] D. J. Griffiths, *Introduction to Quantum Mechanics*, second edition (Pearson Prentice Hall, 2005; Cambridge University Press, 2016).

[104] G. W. Collins, *The Fundamentals of Stellar Astrophysics* (2003) available online at http://ads.harvard.edu/books/1989fsa..book/

[105] Chr. Palmer, E. Loewen, *Diffraction Grating Handbook* (Newport Corporation, 2005).

[106] G. H. Jacoby *et al.*, A Library of Stellar Spectra (1984), available online at: http://cdsarc.u-strasbg.fr/viz-bin/Cat?III/92.

[107] J. Kaler, *STARS*, Information about stars and constellations available online at: http://stars.astro.illinois.edu/sow/sowlist.html

[108] *MILES Spectral Library*, containing ~1000 spectra of reference stars http://miles.iac.es/pages/stellar-libraries/miles-library.php

[109] A. J. Pickles, A Stellar Spectral Flux Library, 1150–25000 Å, *The Publications of the Astronomical Society of the Pacific*, 110, 749, (1998), 863–78, available online at: http://adsabs.harvard.edu/abs/1998PASP..110..863P

[110] *The Bright Star Catalogue*, 5th Revised Edn. Data Access by Alcyone Software: http://www.alcyone.de/search_in_bsc.html

[111] The SAO/NASA Astrophysics Data System, http://adsabs.harvard.edu/index.html

[112] R. Ansgar, Observations of Cool-Star Magnetic Fields (2012), http://solarphysics.livingreviews.org/Articles/lrsp-2012–1/download/lrsp-2012–1Color.pdf

[113] H. W. Babcock, The 34-kilogauss Magnetic Field of HD 215441, *Astrophysical Journal*, 132, (1960), 521, available online at: http://adsabs.harvard.edu/full/1960ApJ...132..521B

[114] F. Borra, J. D. Landstreet, The Magnetic Field Geometry of HD215441 (1977), available online at: http://adsabs.harvard.edu/full/1978ApJ...222..226B

[115] V. G. Elkin *et al.*, A Rival for Babcock's Star: The Extreme 30-kG Variable Magnetic Field in the Ap Star HD75049 (2009), available online at: http://arxiv.org/pdf/0908.0849v2.pdf

[116] J. D. Landstreet, The Measurement of Magnetic Fields in Stars (1979), available online at: http://adsabs.harvard.edu/full/1980AJ.....85..611L

[117] G. Mathys, Ap Stars with Resolved Zeeman Split Lines (1990), available online at: http://articles.adsabs.harvard.edu/full/1990A%26A...232..151M

[118] G. Mathys, T. Lanz, Ap Stars with Resolved Magnetically Split Lines (1991), available online at: http://adsabs.harvard.edu/full/1992A%26A...256..169M

[119] J. P. Aufdenberg *et al.*, The Nature of the Na I D-lines in the Red Rectangle (2011), available online at: http://arxiv.org/pdf/1107.4961.pdf

[120] M. E. Brown *et al.*, A High Resolution Catalogue of Cometary Emission Lines (1996), available online at: http://www.gps.caltech.edu/~mbrown/comet/echelle.html

[121] I. J. Danziger *et al.*, Optical and Radio Studies of SNR in the Local Group Galaxy M33 (1980), available online at: http://www.eso.org/sci/publications/messenger/archive/no.21-sep80/messenger-no21-7-11.pdf

[122] I. J. Danziger *et al.*, Optical Spectra of Supernova Remnants (1975), available online at: http://articles.adsabs.harvard.edu/full/1976PASP...88...44D

[123] K. Davidson, Emission-line Spectra of Condensations in the Crab Nebula (1979), available online at: http://adsabs.harvard.edu/abs/1979ApJ...228..179D

[124] R. J. Glinski *et al.*, On the Red Rectangle Optical Emission Bands (2002), available online at: http://adsabs.harvard.edu/full/2002MNRAS.332L..17G

[125] G. A. Gurzadyan, Excitation Class of Nebulae: an Evolution Criterion? (1990), available online at: http://articles.adsabs.harvard.edu/full/1991Ap%26SS.181...73G

[126] J. Kaler, *The Planetary Nebulae*, available online at: http://stars.astro.illinois.edu/sow/pn.html

[127] L. A. Morgan, The Emission Line Spectrum of the Orion Nebula (1971), available online at: http://articles.adsabs.harvard.edu/cgi-bin/nph-iarticle_query?1971MNRAS.153..393M&defaultprnt=YES&filetype=.pdf

[128] W. A. Reid *et al.*, An Evaluation of the Excitation Parameter for the Central Stars of Planetary Nebulae (2010), available online at: http://arxiv.org/PS_cache/arxiv/pdf/0911/0911.3689v2.pdf

[129] J. D. Thomas, *Spectroscopic Analysis and Modeling of the Red Rectangle*, PhD thesis University of Toledo, (2012), available online at: https://etd.ohiolink.edu/ap/10?0::NO:10:P10_ACCES SION_NUM:toledo1341345222

[130] Y. P. Varshni *et al.*, *Emission Lines Identified in Planetary Nebulae* (2006), available online at: http://laserstars.org/ http://laserstars.org/data/nebula/identification.html

[131] J. R. Walsh *et al.*, Complex Ionized Structure in the Theta-2 Orionis Region, (1981), available online at: http://articles .adsabs.harvard.edu/full/1982MNRAS.201..561W

[132] N. Wehres *et al.*, C2 Emission Features in the Red Rectangle, (2010), available online at: http://adsabs.harvard.edu/abs/ 2010A%26A...518A..36W

[133] Williams College, USA: *Gallery of Planetary Nebula Spectra*, available online at: http://www.williams.edu/astronomy/ research/PN/nebulae/ http://www.williams.edu/astronomy/ research/PN/nebulae/legend.php

[134] D. L. Burke *et al.*, *Precision Determination of Atmospheric Extinction at Optical and Near Infrared Wavelengths* (2010) available online at: http://iopscience.iop.org/0004–637X/720/1/811

[135] L. Colina *et al.*, Absolute Flux Calibrated Spectrum of Vega (1996), available online at: http://www.stsci.edu/instruments/ observatory/PDF/scs8.rev.pdf

[136] *ESO RA Ordered List of Spectrophotometric Standards*, http://www.eso.org/sci/observing/tools/standards/spectra/stanlis .html

[137] J. Rakovský *et al.*, Measurement of Echelle Spectrometer Spectral Response in UV (2011), available online at: http://www .mff.cuni.cz/veda/konference/wds/proc/pdf11/WDS11_245_ f2_Rakovsky.pdf

[138] H. Schmid, Lecture ETHZ, Astronomical Observations Chapter 7: Spectroscopy, available online at: http://www.phys.ethz .ch/~hanschmi/AstObs2012/astobs7.pdf

[139] P. Skoda, Common Methods of Stellar Spectral Analysis and their Support in VO (2011), available online at: http://arxiv .org/abs/1112.2787

[140] W. Stubbs *et al.*, Towards More Precise Survey Photometry for PanSTARRS and LSST: Measuring Directly the Optical Transmission Spectrum of the Atmosphere (2007), available online at: http://arxiv.org/pdf/0708.1364.pdf

[141] R. L. Kurucz *et al.*, The Rotational Velocity and Barium Abundance of Sirius (1977), available online at: http://adsabs .harvard.edu/full/1977ApJ...217..771K

[142] F. Royer, Determination of v sin i with Fourier Transform Techniques (2005), available online at: http://sait.oat.ts.astro .it/MSAIS/8/PDF/124.pdf

[143] Y. Takeda *et al.*, *Rotational Feature of Vega and its Impact on Abundance Determinations* (2007), available online at: http://www.ta3.sk/caosp/Eedition/FullTexts/vol38no2/pp157– 162.pdf

[144] L. Berman, The Effect of Space Reddening on the Balmer Decrement in Planetary Nebulae (1936), available online at: http://adsabs.harvard.edu/full/1936MNRAS..96..890B

[145] G. Burbidge, E. M. Burbidge, Paschen and Balmer Series in Spectra of Chi Ophiuchi and P Cygni (1955), available online at: http://articles.adsabs.harvard.edu/full/1955ApJ...122...89B

[146] G. Burbidge, E. M. Burbidge, The Balmer Decrement in some Be Stars (1953), available online at: http://articles.adsabs .harvard.edu/full/1953ApJ...118..252B

[147] R. Costero, M. Peimbert, The Extinction Law in the Orion Nebula, available online at: http://www.astroscu.unam.mx/ bott/BOTT..5-34/PDF/BOTT..5-34_rcostero.pdf

[148] P. Cox, W. Mathews, Effects of Self-Absorption and Internal Dust on Hydrogen Line Intensities in Gaseous Nebulae (1969), available online at: http://adsabs.harvard.edu/full/ 1969ApJ...155..859C

[149] B. Groves *et al.*, The Balmer Decrement of SDSS Galaxies (2011), available online at: http://arxiv.org/abs/1109.2597

[150] G. Stasinska *et al.* Comparison of Two Methods for Determining the Interstellar Extinction of Planetary Nebulae (1992), available online at: http://articles.adsabs.harvard.edu/full/ 1992A%26A...266..486S

[151] J. R. Ducati *et al.*, Intrinsic Colors of Stars in the Near Infrared (2001) available online at: http://iopscience.iop.org/ 0004–637X/558/1/309/fulltext/

[152] M. Fitzgerald, The Intrinsic Colours of Stars and Two-Colour Reddening Lines (1970), available online at: http://articles .adsabs.harvard.edu/full/1970A%26A.....4..234F

[153] Spectra L200, JTW Astronomy http://www.jtwastronomy .com/products/spectroscopymain.html

[154] M. J. Porter, Spectroscopy on Small Telescopes: The Echelle, (2000), available online at: http://www.kcvs.ca/martin/astro/ spectro/papers/porter.pdf

INDEX

Printed in the United States
By Bookmasters